Introduction to Telecommunications Network Engineering

For a listing of recent titles in the *Artech House Telecommunications Library,*
turn to the back of this book.

Introduction to Telecommunications Network Engineering

Tarmo Anttalainen

Artech House
Boston • London

Library of Congress Cataloging-in-Publication Data
Anttalainen, Tarmo.
Introduction to telecommunications network engineering / Tarmo Anttalainen.
    p. cm. — (Artech House telecommunications library)
  Includes bibliographical references and index.
  ISBN 0-89006-984-0 (alk. paper)
  1. Telecommunication systems. I. Title. II. Series.
TK5105.A55 1998   98-41082
004.6—dc21   CIP

British Library Cataloguing in Publication Data
Anttalainen, Tarmo
  Introduction to telecommunications network engineering
  1. Telecommunication systems 2. Telecommunication systems—
Handbooks, manuals, etc.
  I. Title
  621.3'82

  ISBN 0-89006-984-0

Cover design by Lynda Fishbourne

© 1999 ARTECH HOUSE, INC.
685 Canton Street
Norwood, MA 02062

International Standard Book Number: 0-89006-984-0
Cataloging-In-Publication: 98-41082

10 9 8 7 6 5 4 3

Contents

Preface

Telecommunications is one of the fastest growing business sectors of modern Information Technologies. A couple of decades ago, in order to have a basic understanding of telecommunications, it was enough to know the operation of the telephone network. Today telecommunications include a vast variety of modern technologies and services. Some services, such as the fixed telephone service in developed countries, are becoming mature; and some are exploding (e.g., cellular mobile communications). The deregulation of the telecommunications business has accelerated business growth even though, or maybe because, tariffs have decreased.

The present telecommunications environment, in which each of us has to make choices, has become complicated. It contains multiple operators that provide a continuously changing variety of services. Telecommunications has also become a strategically important resource for most modern corporations, and its importance is still increasing. Special attention has to be paid to the security aspects and cost of telecommunications. The new environment provides new options for users, and we should be more aware of telecommunications as a whole in order to be able to utilize the possibilities available today.

The business of telecommunications is growing rapidly, and many more newcomers are finding employment in this area. Even if they may have a technical background, they often feel that they have a very restricted overall view of the telecommunications network as a whole. The first purpose of this book is to give a wide view of telecommunications networks to newcomers entering the telecommunications business. This kind of general knowledge is useful for the users of telecommunication services, the operator personnel, as well as the employees of telecommunication system manufacturers.

The professionals working with these complicated technologies often have a very deep knowledge of one very narrow section of telecommunications but are not familiar with the hundreds of terms and abbreviations that are used in other telecommunication areas by individuals with whom they need to cooperate. One purpose of this book is to provide definitions for some of the most common terms and abbreviations used in different areas of telecommunications.

For the eighteen years during which I worked in system development activities for Nokia Telecommunications I noticed that there are relatively few books that give a good introduction to data, fixed, and mobile networks. In the modern business environment, all people—such as the development engineers, testing personnel, and sales managers—must have a common language in order to work together efficiently. Currently, there are no books that give an overview of telecommunications as a whole. Most of the books on the market explain telecommunications from only one point of view even though there is no longer any distinct separation of networks providing data, speech, and mobile services. Some sections of this book were originally written for a basic course on telecommunications networks for new employees at Nokia Telecommunications.

Presently, the material included in this book is used in the Telecommunications Networks course for students of information technologies at the Espoo-Vantaa Institute of Technology in Finland. This course aims to give a basic understanding of the structure and operation of a telecommunications network. A deeper theory of telecommunications, such as the spectral analysis of signals or the detailed knowledge of the operation and functions of data and mobile networks is given in dedicated courses. These subjects are only briefly introduced in this book.

However, I have tried not to cover too many aspects of modern telecommunications in this book in order to keep its structure clear. The aim is to lay the groundwork for later studies of telecommunications, for which there are many good sources available. Some of them are listed at the end of each chapter.

Objectives

This book aims to give answers to the fundamental questions concerning telecommunications networks and services, telecommunications as a business area, and the general trends of technical development. These questions include:

- What is the structure and what are the main components of a modern telecommunications network?

- What are the main reasons for the growth of telecommunications as a business area?
- What is the importance of standardization, and what are the main standardization bodies for telecommunications?
- Why do we prefer the modern digital technologies rather than the analog technologies of the past?
- How are analog signals processed for transmission over digital circuits?
- What are the basic techniques used in a primary pulse code modulation system that transmits analog speech through the digital telecommunications network?
- What are the fundamental differences between circuit and packet switching techniques?
- What are modems, baseband modems, local area networks, and leased line circuits; and what is meant by asynchronous transfer mode?
- How does the digital Integrated Services Digital Network differ from the ordinary telephone network?
- What are the fundamental limiting factors of the rate of information transmission through a transmission channel?
- How do cellular mobile networks operate, and what are their main components?
- What seem to be the major developments in telecommunications in a few years to come?

For those readers who hope to use this book as a training tool, slides of the figures and solutions to the problems are available from the author. The figures of this book include some key points written in text in order to make them useful for lecture material. Figures and solutions of the problems are available in electronic and paper format. If you have a need for them, please contact the author at tarmo.anttalainen@evitech.fi.

Acknowledgments
I thank my wife Pirjo and my children Heini, Sini, and Joni for enduring those lengthened working days and missed weekends while I drafted and redrafted the manuscript. I am indebted to my friend Kyosti Anttonen for his review and valuable proposals in the development of the material. I also thank my students, especially Mr. Esko Halttunen, for their helpful contributions, and Espoo-Vantaa Institute of Technology for the opportunity to complete this project.

1

Introduction to Telecommunications

1.1 What is Telecommunications?

Telecommunications has been defined as a technology concerned with communicating from a distance, and we can categorize it in various ways. Figure 1.1 shows one possible division. It includes mechanical communication and electrical communication because telecommunications has developed from mechanical to electrical using increasingly more sophisticated electrical systems. This is why many authorities such as national *Post, Telegraph, and Telephone* (PTTs) are involved in telecommunications by both means.

Our main concern here is electrical and bidirectional communication, as shown in the upper part of Figure 1.1. The share of the mechanical telecommunications such as conventional mail and press is expected to decrease; while the electrical, especially bidirectional, communication will increase and take the major share of overall turnover of telecommunications in the future. Hence, major press corporations are interested in electrical telecommunications as a business opportunity. Telecommunication is expected to be one of the most rapidly growing business sectors during the next few years.

1.2 Significance of Telecommunications

Many different telecommunications networks are interconnected into a continuously changing and extremely complicated global system. We look at telecommunications from different points of view in order to understand how complicated a system we are dealing with and how dependent we are on it.

1

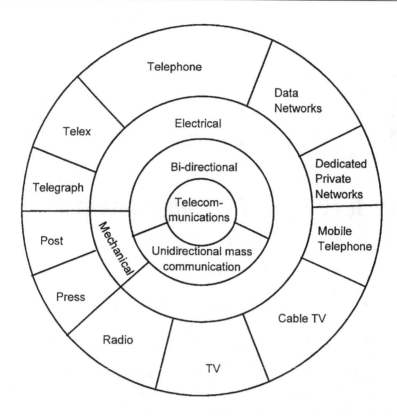

Figure 1.1 Telecommunications.

Telecommunications networks have the most complicated equipment in the world.

Let us think only about the telephone network, which includes approximately 800 million telephones with universal access. When any of these telephones requests a call, the telephone network is able to establish a connection to any other telephone in the world. In addition, there are many other networks that are interconnected with the telephone network. This gives us a view of the complexity of the global telecommunications network; there is no other system that exceeds the complexity of telecommunications networks in the world.

Telecommunication services have an essential impact on the development of a community.

If we look at the telephone density of a country we can estimate its level of technical and economical development. In developing countries the telephone density is less than 10 telephones per 1000 inhabitants; in developed countries, like North America and Europe, there are 500 to 600 telephones per 1000 inhabitants. The economic development of developing countries depends on (in addition to many other things) the availability of telecommunication services.

The operations of a modern community are highly dependent on telecommunications.

We can hardly imagine our working environment without telecommunication services. The *local area network* (LAN), to which our computer is connected is interconnected with the LANs of other sites of the company. This is mandatory so that departments can work together efficiently. We communicate daily with people in other organizations with the help of electronic mail, telephone, facsimile, and mobile telephone. Governmental organizations that provide public services are as dependent on telecommunication services as private organizations.

Telecommunications have an essential role in many areas of everyday living.

Everyday life is dependent on telecommunications. Each of us uses telecommunication services and services that rely on telecommunications daily. Services that are dependent on telecommunications include:

- Banking, automatic teller machines, telebanking;
- Aviation, booking of tickets;
- Sales, wholesale, and order handling;
- Credit card payments at gasoline stations;
- Booking hotel rooms by travel agencies;
- Material purchasing by industry;
- Government operations, such as taxation.

1.3 Historical Perspective

Some of the most important milestones in the development of electrical telecommunication systems according to [1] are indicated in this section. Terms and abbreviations in the chronology will be explained in later chapters of this book. The development and expansion of some telecommunication services is also illustrated in Figure 1.2.

1800–1837 *Preliminary developments*: Volta discovers the primary battery; Fourier and Laplace present mathematical treatises; Ampere, Faraday, and Henry conduct experiments on electricity and magnetism; Ohm's law (1826); Gauss, Weber, and Wheatstone develop early telegraph systems.

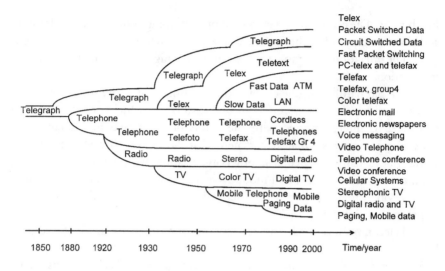

Telex
Packet Switched Data
Circuit Switched Data
Fast Packet Switching
PC-telex and telefax
Telefax
Telefax, group4
Color telefax
Electronic mail
Electronic newspapers
Voice messaging
Video Telephone
Telephone conference
Video conference
Cellular Systems
Stereophonic TV
Digital radio and TV
Paging, Mobile data

Telegraph
Teletext
Fast Data ATM
Slow Data LAN
Telegraph Telex
Telegraph Telex
Telegraph
Telephone
Telephone Telephone Cordless
Telephone Telefoto Telefax Telephones
Telefax Gr 4
Radio Radio Stereo Digital radio
TV Color TV Digital TV
Mobile Telephone Mobile
Paging Data

1850 1880 1920 1930 1950 1970 1990 2000 Time/year

Figure 1.2 Development of telecommunications services and services.

1838–1866 *Telegraphy*: Morse perfects his system; Steinhill finds that the Earth can be used for a current path; commercial service is initiated (1844); multiplexing techniques are devised; William Thomson calculates the pulse response of telegraph line (1855); transatlantic cables are installed.

1845 Kirchoff's circuit laws.

1864 Maxwell's equations predict electromagnetic radiation.

1876–1899 *Telephony*: Alexander Graham Bell perfects the acoustic transducer; first telephony exchange with eight lines; Edison's carbon-button transducer; cable circuits are introduced; Strowger devises automatic step-by-step switching (1887); Pupin presents the theory of loading.

1887–1907 *Wireless telegraphy*: Heinrich Hertz verifies Maxwell's theory; demonstrations by Marconi and Popov; Marconi patents complete wireless telegraph system (1897); commercial service begins, including ship-to-shore and transatlantic systems.

1904–1920 *Communication electronics*: Lee De Forest invents the Audion (triode) based on Fleming's diode; basic filter types are devised; experiments with AM radio broadcasting; the Bell System completes the transcontinental telephone line with electronic repeaters (1915); multiplexed carrier telephony is introduced: H. C. Armstrong perfects the superheterodyne radio receiver (1918); first commercial broadcasting station.

1920–1928 Carson, Nyquist, Johnson, and Hartley present their transmission theory.

1923–1938 *Television*: Mechanical image-formation system is demonstrated; theoretical analysis of bandwidth requirements; DuMont and others perfect vacuum cathode-ray tubes; field tests and experimental broadcasting begin.

1931 Teletypewriter service initiated.

1934 H. S. Black develops the negative feedback amplifier.

1936 Armstrong's paper states the case of FM radio.

1937 Alec Reeves conceives pulse code modulation.

1938–1945 Radar and microwave systems are developed during World War II; FM is used extensively for military communications; hardware, electronics, and theory are improved in all areas.

1944–1947 Mathematical representations of noise are developed; statistical methods for signal detection are developed.

1948–1950 C. E. Shannon publishes the founding papers of information theory; Hamming and Golay devise error-correcting codes.

1948–1951 Transistor devices are invented.

1950 Time division multiplexing is applied to telephony. Hamming presents the first error-correction codes.

1953 Color TV standards are established in United States.

1955 J. R. Pierce proposes satellite communication systems.

1958 Long-distance data transmission system is developed for military purposes.

1960 Maiman demonstrates the first laser.

1961 Integrated circuits are applied to commercial production.

1962 Satellite communication begins with Telstar I.

1962–1966 Data transmission service is offered commercially; wideband channels are designed for digital signaling; *pulse code modulating* (PCM) proves feasible for voice and TV transmission; theory for digital transmission is developed; Viterbi presents error-correcting codes; adaptive equalization is developed.

1964 Fully electronic telephone switching system is put into service.

1965 Mariner IV transmits pictures from Mars to Earth.

1966–1975 Commercial satellite relay becomes available; optical links using lasers and fiber optics; ARPANET is created (1969) and followed by international computer networks.

1968–1969 Digitalization of telephone network begins.

1970–1975 Standards of PCM by CCITT are developed.

1975–1985   High-capacity optical systems are developed; the breakthrough of optical technology and fully integrated switching systems; digital signal processing by microprocessors.

1980–1985   Modern cellular mobile network is put into service: NMT in Northern Europe and AMPS in the United States; OSI reference model is defined by *International Standards Organization* (ISO).

1985–1990   LANs breakthrough; *Integrated Services Digital Network* (ISDN) standardization finalized; public data communication services become widely available; Optical transmission systems replace copper systems in long-distance wideband transmission; SONET is developed; *Global System for Mobile* (GSM) and SDH standardization is finalized and first systems put into commercial use.

1990–1997   The first digital cellular system GSM is put into commercial use and its breakthrough is felt worldwide; deregulation of telecommunications in Europe proceeds and satellite-TV systems become popular; Internet usage and services expand rapidly because of the WWW.

1997–2001   Telecommunication community is fully deregulated and business grows rapidly; cellular networks such GSM and CDMA expand worldwide; Internet traffic exceeds *packet-switched traffic network* (PSTN) traffic; commercial applications of Internet expand and a share of conventional speech communications is transferred from PSTN to Internet; ATM technology makes *wide area networks* (WAN) networks wideband; performance of LANs improve with Gbps technologies.

2001–   *High-definition TV* (HDTV) and integrated new third-generation mobile communication systems are put into use; broadband networks and access systems bring new multimedia services available to everybody.

1.4 Standardization

Communication networks are designed to serve a wide variety of users with equipment from many different vendors. To design and build networks effectively, standards are necessary to achieve interoperability, compatibility and required performance in a cost-effective manner.

Standards (open standards) are needed to enable the interconnections of systems, equipment and networks of different manufacturers, vendors, and operators. The most important advantages and some other aspects of open telecommunication standards are explained.

Standards enable competition.

Open standards are available to any telecommunication system vendor. When a new system is standardized and if it is attractive from a business point of view, multiple vendors appear in this new market. As long as a system is proprietary, specifications are the property of one manufacturer and it is difficult, if not impossible, for a new manufacturer to start to produce compatible competing systems. Open competition makes products more cost effective, therefore providing low-cost services to telecommunication users.

Standards lead to economies of scale in manufacturing and engineering.

Standards increase the market for products adhering to the standard, which leads to mass production and economies of scale in manufacturing and engineering, *very large scale integration* (VLSI) implementations, and other benefits that decrease price and further increase the acceptance. This supports the economic development of the community by improving telecommunication services and decreasing their cost.

Political interests often lead to different standards in Europe, Japan, and America.

Standardization is not only a technical matter. Sometimes opposing political interests make the approval of global standards impossible and different standards are often adapted for Europe, America, and Japan. Europe does not want to accept American technology and America does not want to accept the European one in order to protect local industry. One example of a political decision was to define (in the seventies) a different PCM coding law for Europe instead of the existing American PCM code. (We will explain these in Chapter 3.) A more recent example is the American decision not to accept European GSM technology as a major digital cellular technology.

International standards are threats to the local industries of big countries but opportunities to the industries of small countries.

Major manufacturers of big countries may not support international standardization because it would open their local market to international competition. Manufacturers of small countries strongly support global standardization because they are dependent on foreign markets. Their home market is not large enough and they are looking for new markets for their technology.

Standards make possible the interconnection of systems from different vendors possible.

The main technological aim of standardization is to make systems in the network or different networks "understand" each other. Technical specifications included in standards make systems compatible and support the provision of wide area or even global services that are based on standardized technology.

Standards make users and network operators vendor independent and improve availability of the systems.

A standardized interface between terminal and network enables subscribers to purchase terminal equipment from multiple vendors. Standardized interfaces between systems in the network enable network operators to use multiple competing suppliers of systems. Standardization improves the availability and quality of systems and reduces their cost.

Standards make international services available.

Standardization plays a key role in providing international services. Official global standards define, for example, telephone service, ISDN, global X.25 packet-switched data service, telex, and facsimile. Standards of some systems may not have an official worldwide acceptance; but if the system becomes popular all around the world, a worldwide service may become available. Recent examples of these services are GSM communications and Internet with WWW.

Some examples of international standardization areas clarify the influence of standardization on our everyday living include:

- Screw thread pitches (ISO, Technical Committee 1): one of the first activity areas of standardization. In the 1960s a bolt from one car did not fit another. Currently they are internationally standardized and most often compatible.

- International telephone numbering, country codes: without globally unique identification of subscribers, automatic international telephone calls would not be available.

- Telephone subscriber interface.

- PCM coding and primary rate frame structure: make national and international digital connections between networks possible.

- Television and radio systems.

- Frequencies used for satellite and other radio communications.

- Connectors and signals of PC, printer, and modem interfaces.

- *LANs*: enable us to use computers from any manufacturer in our company network, for example.

1.5 Standards Organizations

There are many organizations involved in standardization work. We look at them from two points of view: who are the players of telecommunication business involved in standardization and which are the authorities that approve official standards?

1.5.1 Interested Parties

Figure 1.3 shows some groups that are interested in standardization and participate in standardization work. We list these parties and their most important
interests, that is, why they are involved in standardization work.

Network operators support standardization:

- To improve the compatibility of telecommunication systems;
- To be able to provide wide-area or even international services;
- To be able to purchase equipment from multiple vendors.

Equipment manufacturers participate in standardization:

- To get information about future standards for their development activi-
  ties as early as possible;
- To support standards that are based on their own technologies;
- To prevent standardization if it opens their own markets.

Service users participate in standardization:

- To support the development of standardized international services;
- To get alternative system vendors (multivendor networks);
- To improve the compatibility of their systems.

Other interested parties include:

- Government officials who are keen on having national approaches
  adopted as international standards;
- Academic experts who want to become inventors of new technological
  approaches.

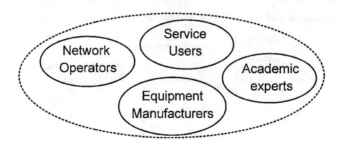

Figure 1.3 Interested parties.

1.5.2 National Standardization Authorities

National standardization authorities approve official national standards. Many international standards include alternatives, and options, from which national authorities selects one for a national standard. These options are included because a common global understanding was not found. Sometimes some aspects are left open and they require a national standard. For example national authorities define the national telephone numbering plan and frequency allocation inside their country. Some examples of national authorities are shown in the Figure 1.4. They take care of all the standardization areas, and they have specialized organizations or working groups for the standardization of each specific technical area like telecommunications and information technology.

1.5.3 European Organizations

The most important European standards organizations are shown in Figure 1.5. They are responsible for developing Europe-wide standards to open national borders in order to improve Pan-European telecommunication services.

The *European Telecommunications Standards Institute* (ETSI) is an independent body for making standards for the European Community. Telecommunication network operators and manufacturers participate in standardization work.

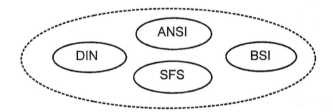

Figure 1.4 Some examples of national standardization authorities (BTI, British Standards Institute; DIN, Deutche Industrie-Normen; ANSI, American National Standards Institute; SFS, Finnish Standards Institute).

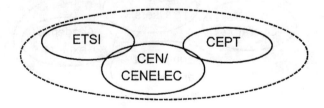

Figure 1.5 European standards organizations.

The *European Committee for Electrotechnical Standardization/European Committee for Standardization* (CEN/CENELEC) is a joint organization for the standardization of information technology. It corresponds to IEC/ISO on a global level and takes care of environmental and electromechanical aspects.

The *Conférence Européenne des Administrations des Postes et des Telecommunications* or *European Conference of Posts and Telecommunications Administrations* (CEPT) was doing the work of ETSI before the European Commission Green Paper opened competition in Europe within the telecommunications market. The deregulation of telecommunications required national PTTs to become network operators equal to other new operators and they are not allowed to make standards any more.

One example of standards made by ETSI is the digital cellular mobile system GSM that is accepted, in European countries and as a major standard for digital mobile communications worldwide.

1.5.4 American Organizations

The national standards authority, the *American National Standards Institute* (ANSI) of the United States has accredited several organizations to work for standards for telecommunications. Some of these organizations are shown in Figure 1.6.

The *Institute of Electrical and Electronics Engineers* (IEEE) is one of the largest professional societies in the world and has produced many important standards for telecommunications. Some of these standards, such as the standards for LANs, have been accepted by the ISO as international standards. Another example is the international standard for "Ethernet": ISO 8002 is presently equal to the IEEE 802.2 standard.

The *Electronic Industries Association* (EIA) is an American organization of electronic equipment manufacturers. Many of its standards, such as the connectors of personal computers, have achieved global acceptance. For example, the data interface standard EIA RS-232 is compatible with the V.24/28 recommendations of ITU-T.

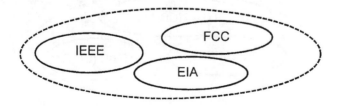

Figure 1.6 American standards organization.

The *Federal Communications Commission* (FCC) is not actually a standards body but a regulatory body. It is a government organization that regulates wire and radio communications. It has played an important role, for example, in the development of worldwide specifications for radiation and susceptibility of electromagnetic disturbances of telecommunication equipment.

1.5.5 Global Organizations

The *International Telecommunications Union* (ITU) is a specialized agency of the United Nations responsible for telecommunications. It has nearly two hundred member countries, and standardization work is divided between two major standardization bodies: the ITU-T (previously known as CCITT) and ITU-R (previously known as CCIR); see Figure 1.7.

The *Comité Consultatif International de Télégraphique et Téléphonique*, or *International Telegraph and Telephone Consultative Committee* (CCITT/ITU-T), is presently called ITU-T, where "T" stands for telecommunications.

The *Comité Consultatif International des Radiocommunications*, or *International Radio Consultative Committee* (CCIR/ITU-R), is presently known as ITU-R, where "R" stands for radio.

ITU-T and ITU-R publish recommendations that are in fact strong standards for telecommunications networks. ITU-T works for the standards of public telecommunication networks (e.g., ISDN), and ITU-R works with radio aspects like the usage of radio frequencies worldwide and specifications for radio systems. Many parties participate in their work, but only national authorities may vote. ITU-T, formerly CCITT, has created most of the current worldwide standards for public networks.

The *International Standards Organization/International Electrotechnical Commission* (ISO/IEC) is a joint organization responsible for the standardization of information technology.

ISO has done important work in the area of data communications and protocols and IEC in the area of electromechanical (e.g., connectors) and environmental aspects.

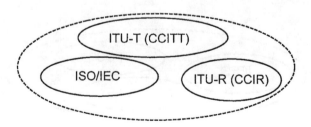

Figure 1.7 Global standards organizations.

1.5.6 Other Organizations

There are many other organizations working with standards. Some of these are active in ITU-T and ISO, and many international standards are based on (or may be even copies of) the initial work of these groups. We introduce some of these as examples of standards organizations without official status; see Figure 1.8.

The *Internet Engineering Task Force* (IETF) takes care of the standardization of the TCP/IP protocol suite used in the Internet.

The *Asynchronous Transfer Mode Forum* (ATM forum) is an open organization of ATM equipment manufacturers that support compatibility between systems from different vendors. It is more flexible and can produce necessary standards on a shorter time scale than official global organizations. Their standards are often used as a basis for official standards later approved by ITU-T and ISO.

The *Network Management Forum* is an organization of system manufacturers that works to speed up the development of management standards. With the help of these standards telecommunication network operators should be able to control and supervise their multivendor networks efficiently from the same management center. Proposals are then given to ITU-T and ISO for official international acceptance.

There are many other such organizations; new groups appear and some organizations disappear every year. Examples of current active groups are *Global System for Mobile Communications* (GSM), *Memorandum of Understanding* (MoU) and *Asymmetric Digital Subscriber Line Forum* (ADSL Forum).

1.6 Development of Telecommunication Business

In the past, telecommunications was a protected business area. The national PTTs were once the only national telecommunication operators in most countries. They had control over standardization in international standardization

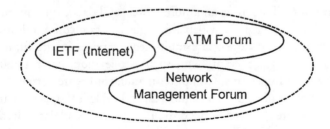

Figure 1.8 Other standardization organizations.

bodies and a monopoly in providing telecommunication service in their home country. For political reasons domestic manufacturers were preferred as suppliers of the systems needed in the network. Competition was not allowed, and the development of services and networks was slow in many countries.

During the latter part of the 1980s the deregulation of the telecommunication business started in Europe; it is presently proceeding rapidly in many other areas of the world as well. Competitive telecommunication services are important for the development of an economy, and this development for free markets is heavily supported by governments. In Europe the European Union has paid much attention to the deregulation of the telecommunication business. New operators have obtained licenses to provide local and long-distance telephone and data services and mobile telecommunication services. Previously many standards, such as analog mobile telephone standards, did not even support a multioperator environment. The initial requirement of the digital mobile telecommunication system (GSM) in Europe was the support of multiple networks in the same geographical area. This deregulation of telecommunication business has reduced tariffs for long-distance calls and mobile calls to a small fraction of the tariffs in the mid-1980s. The reduction of fees has further increased the demand for services, which has prompted reductions in the price of terminal equipment, such as mobile telephones, and the fees for calls.

This development has also shown how dangerous it is for manufacturers to be too dependent on a single domestic customer. Many telecommunication manufacturers that were independent ten years ago do not exist as independent suppliers anymore. This process will continue. At the same time new small manufacturers are appearing. Their window of opportunity is to produce special equipment, in which the largest vendors are not interested, or systems for brand-new rapidly growing services.

The *Plain Old Telephone Service* (POTS) will still be important in the future; and new technologies, such as *Wireless Local Loop* (WLL) or *Radio Local Loop* (RLL), will help new operators compete in providing even local fixed telephone service. However, mobile and data communications grow most rapidly in volume. The two main directions of this development are that voice communications will be more and more mobile and data communications will develop into more and more wideband, high data rate communications.

The provision of developing multimedia services in the future will be especially interesting. Entertainment, such as television, has previously been provided by special operators. Computer manufacturers are currently involved in the development of future TV-standards, and cable-TV operators and newspaper publishers are involved in the provision of speech and data telecommunication services via cable-TV networks. The expansion of the Internet, with its improving capability to transmit voice in addition to data, is a new challenge to public network operators. In the next few years we will see many different business structures among operators. Telecommunication operators as well as manufacturers have to develop ways to integrate their infrastructure to support the provision of many different mixtures of services.

References

[1]   Carlsson, A. B., Communication Systems, *An Introduction to Signals and Noise in Electrical Communication,* New York, NY: McGraw-Hill Book Company, 1986.

2

The Telecommunications Network: An Overview

This chapter describes the basic operation of a telecommunications network with the help of a conventional telephone. The operation of a conventional telephone, which is easy to understand, is used to clarify how telephone connections are built up in the network. We look at the subscribers signaling over the subscriber loop of the telephone network. The same main signaling phases are needed in modern data and mobile networks. We start with this simple service to lay a foundation for understanding more complicated services in later chapters. In this chapter we divide the network into layers and briefly describe different network technologies that are needed to provide various kinds of services. Some of these, such as mobile and data networks and their services, are discussed in more detail later in this book. The last topic in this chapter is an introduction to the theory of traffic engineering, that is, how much capacity we should build into the network in order to provide a sufficient grade of service for the customers.

2.1 Basic Telecommunications Network

The basic purpose of a telecommunications network is to transmit user information in any form to another user of the network. These users of public networks, such as a telephone network, are called subscribers. User information may have many forms, such as voice or data, and subscribers may use different access network technologies to access the network, such as fixed or cellular telephones. We shall see that a telecommunications network consists of many different networks providing different services, for example, data, fixed, or cellular telephony service. These different networks are discussed in later chapters. In the following section we introduce the basic functions that are needed in any networks regardless of what services they provide.

The three technologies needed for communication through a network are:

- Transmission;
- Switching;
- Signaling.

Each of these technologies requires specialists for engineering, operation, and maintenance.

2.1.1   Transmission

Transmission is the process of transporting information between end points of a system or a network. Transmission systems use three basic media for information transfer from one point to another:

- Copper cables, such as LANs and telephone subscriber lines;
- Optical fiber cables, such as high data-rate transmission in a telecommunications network;
- Radio waves, such as cellular phones and satellite transmission.

In a telecommunications network the transmission systems interconnect exchanges, and these transmission systems altogether are called the *Transmission* or *Transport Network*. Note that the number of speech channels (which is one measure of transmission capacity) needed between exchanges is much smaller than the number of subscribers since only a small fraction of them has a call connected at the same time. In Chapter 4 we will discuss transmission in more detail.

2.1.2   Switching

In principle all telephones could still be connected by cables as they were in the very beginning of the history of telephony. However, when the number of telephones grew, it was soon noticed that it was necessary to switch signals from one wire to another. Then only a few cable connections were needed between exchanges because the number of ongoing calls was much smaller than the number of telephones; see Figure 2.1. The first switches were not automatic and switching was done manually on switchboards.

Automatic switches, known as exchanges, were developed in 1887 by Strowger. Then the switching had to be controlled by the telephone user with the help of pulses generated by a dial. For many decades exchanges were a complex series of electromechanical selectors, but during the last twenty years they have developed into software-controlled digital exchanges that can provide additional services. Modern exchanges usually have quite a large capacity, tens of thousands subscribers, and thousands of them may have an ongoing call at the same time.

2.1.3 Signaling

Signaling is the mechanism that allows network entities (customer premises or network switches) to establish, maintain, and terminate sessions in a network. Signaling is carried out with the help of specific signals or messages that indicate to the other end what is requested of it by this connection. Some examples about signaling examples on subscriber lines are:

- *Off-hook condition*: the exchange notices that the subscriber has raised the telephone hook (DC-loop is broken) and gives a dial tone to the subscribers.

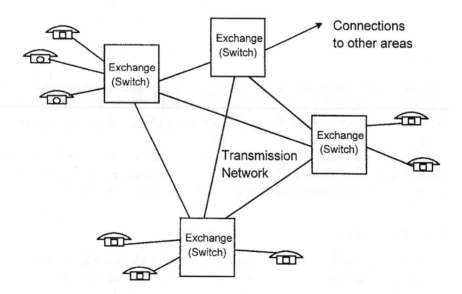

Figure 2.1 A basic telecommunications network.

- *Dial*: the subscriber dials digits and that are received by the exchange.

- *On-hook condition*: the exchange notices that the subscriber has finished the call (subscriber loop is connected), clears the connection, and stops billing.

Signaling is naturally needed between exchanges as well because most calls have to be connected via more than one exchange. Many different signaling systems are in use for the interconnection of different exchanges. Signaling is an extremely complex matter in a telecommunications network. Imagine, for example, a foreign GSM subscriber switching his telephone on in Hong Kong. In a few seconds he is able to receive calls directed to him. Information transferred for this function is carried in hundreds of signaling messages between exchanges in international and national networks. Signaling in the subscriber loop is discussed in Section 2.3 and signaling between exchanges in Section 2.6.

2.2 Operation of a Conventional Telephone

An ordinary home telephone receives the electrical power that it needs for operation from the local exchange via two copper wires. These two wires that carry a speech signal as well are called a "pair" or a "local loop." This principle of power supply makes basic telephone service independent from an electric power network. Local exchanges have a large-capacity battery that keeps the exchange and subscriber sets operational for a few hours if the supply of electricity is cut off. This is essential because the operation of the telephone network is especially important in emergency situations when the electric power supply may be down. Figure 2.2 shows a simplified figure of the telephone connection. Elements of the figure and operation of the subscriber loop are explained later in this chapter.

Minor operational differences, particularly in the provision of PBX/PABX systems, exist around the world, but the principles discussed in this chapter apply to the overwhelming majority of PSTN systems.

2.2.1 Microphone

When we raise the hook of a telephone, the on/off hook switch is closed and current starts flowing on the subscriber loop through the microphone that is connected to the subscriber loop. The microphone converts acoustic energy to electrical energy. Originally telephone microphones were so-called carbon

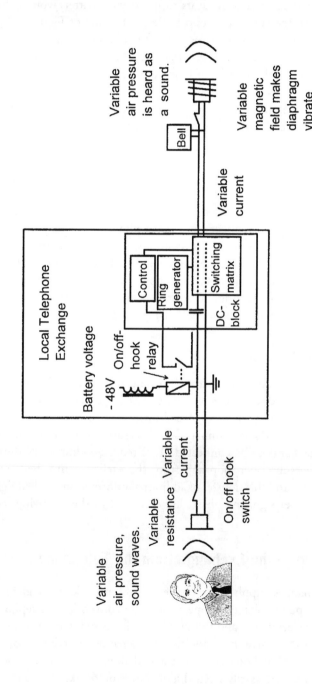

Figure 2.2 Operation principle of a conventional telephone.

microphones that had diaphragms with small containers with carbon grains and they operated as variable resistors supplied with battery voltage from an exchange site; see the subscriber loop on left-hand side of Figure 2.2. When a sound wave presses carbon grains more tightly, loop resistance decreases and current slightly increases. The variable air pressure generates a variable, alternating current to the subscriber loop. This variable current contains voice information.

The basic operation principle of the subscriber loop is still the same, although modern telephones include more sophisticated and better quality microphones.

2.2.2 Earphone

Alternating current, generated by the microphone, is converted back into voice at the other end of the connection. The earphone has a diaphragm with a piece of magnet inside a coil. The coil is supplied by alternating current produced by the microphone at the remote end of the connection. The current generates a variable magnetic field that moves the diaphragm that produces sound waves close to the original sound at the transmitting end (see the subscriber loop on the right-hand side of Figure 2.2).

2.2.3 Signaling Functions

The microphone generates the electrical current that carries voice information, and the earphone produces the voice at the receiving end of the speech circuit. The telephone network provides a dialed-up or switched service that enables the subscriber to initiate and terminate calls. The subscriber dials up the number to which he/she wants to be connected. This requires additional information transfer over the subscriber loop and from the exchange to other exchanges on the connection, and this transfer of additional information is called signaling. The basic subscriber signaling phases are described in the following section.

2.3 Signaling to the Exchange from the Telephone

Telephone exchanges supply DC-voltage to subscriber loops, and telephone sets use this supplied voltage for operation. The conventional telephone does not include any electronics, and the supplied voltage and current were directly used for speech transmission in addition to signaling functions that include the detection of on/off hook condition and dialing. Modern electronic telephones would not necessarily need this if they would take their power from a power socket at home. However, the power supply from the exchange is still an important feature. It ensures that the telephone network operates even in emergency situations when the electrical power network may be down.

2.3.1 Set-Up and Release of a Call

Each telephone has a switch that indicates an on- or off-hook condition. When the hook is raised, the switch is closed and an approximately 50-mA current starts flowing. This is detected by a relay giving information to the control unit in the exchange; see Figure 2.2. The control unit is an efficient and reliable computer in the telephone exchange. It activates signaling circuits that then receive dialed digits from the A-subscriber (we call a subscriber who initiates a call an A-subscriber and a subscriber who receives a call a B-subscriber). The control unit in the telephone exchange controls the switching matrix that connects the speech circuit through to the called B-subscriber. A connection is made according to the numbers dialed by the A-subscriber.

When a call is coming to the B-subscriber, the telephone exchange supplies a ringing voltage to the subscriber loop and the bell of the telephone starts ringing. The ringing voltage is often about 70V AC with 25-Hz frequency, which is high enough to activate the bell on any telephone. The ringing voltage is switched off immediately when an off-hook condition is detected on the loop of subscriber B, and then an end-to-end speech circuit is connected and the conversation may start.

Figure 2.3 shows the signaling phases on a subscriber loop. When the exchange detects the off-hook condition of a subscriber loop, it informs us with a dial tone that we hear when we raise the hook that it is ready to receive digits. After dialing it keeps us informed whether the circuit establishment is successful by sending us a ringing tone when the telephone at the other end rings. When the B-subscriber answers, the exchange switches off both the ringing signal and the ringing tone and connects the circuit through. At the end of the conversation an on-hook condition is detected by the exchange and the speech circuit is released.

In the next sections we explain in more detail one of the subscriber signaling phases, and the transmission of dialed digits from a subscriber's telephone to the local exchange.

2.3.2 Rotary Dialing

There is a switch in the telephone set that is opened for an on-hook condition and closed when the hook is off. This indicates to the telephone exchange when there is a call to be initiated and when it has to prepare to receive dialed

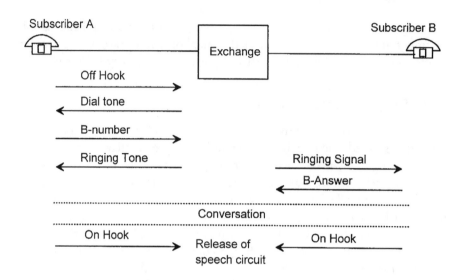

Figure 2.3 Subscriber signaling.

digits. In old telephones that exchanges still have to support, this method of local-loop connection/disconnection is used to transmit dialed digits as well; see Figure 2.4. We call this principle rotary or pulse dialing.

In rotary dialing a local loop is closed and opened according to the dialed digits and the number of current pulses is detected by the local exchange. This signaling methods is also known as loop disconnect signaling. The main disadvantages are that it is slow and expansive due to the high-resolution mechanics and it cannot support new services such as call forwarding. Local-loop interfaces of the telephone exchanges have to support this technology though it has been gradually replaced by tone dialing.

When a digit is to be dialed, the dialing plate with finger holes is rotated clockwise to the end and released. While homing, the switch is breaking the line current periodically and the number of these periods indicates the dialed digit. For example, digit 1 has one period, 2 has two periods, and 0 has ten

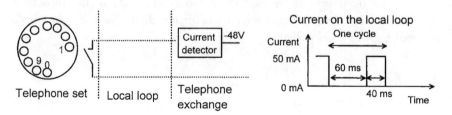

Figure 2.4 Rotary/pulse dialing.

periods or cycles. Mechanics make the homing speed approximately constant and each period is about 100-ms long with a 60-ms break; see Figure 2.4. This method for the transmission of digits is sometimes also used for signaling between exchanges, for which it is known as "Loop Disconnect Signaling."

The value of the loop current differs slightly from country to country and is also dependent on line length and supply voltage, for example. Typically it is 20 mA to 50 mA, which was high enough to control old-generation electromechanical switches that used pulses to directly control rotating switches of the switching matrix of an exchange.

2.3.3 Tone Dialing

Modern telephones include electronic circuits that make the implementation of better means for signaling feasible. Digital exchanges do not require high-power pulses to drive the selectors like old electromechanical switches did. However, subscriber lines are still, and will be, supplied by a −48- or −60-V battery to make operation of telephones independent from the electric power supply. Electronic telephones use 50- to 500-μA current all the time to supply power to their electronic circuitry, which is needed for number repetition, abbreviated dialing, and other additional features of modern telephone sets.

In modern telephones there are usually 12 push buttons (keys A to D of Figure 2.5 are not included in an ordinary subscriber set) for dialing, each generating a tone with two frequencies. One of the frequencies is from the upper frequency band and the other from the lower band. All frequencies are inside the voice frequency band (300 Hz to 3400 Hz) and can thus be transmitted through the network from end to end when the speech connection is established. This signaling principle is known as *dual-tone multifrequency* (DTMF) signaling.

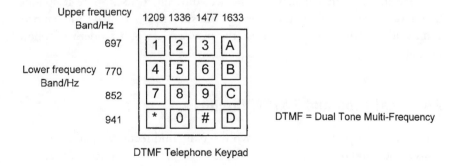

Figure 2.5 Tone dialing. Note that push buttons A, B, C, and D are not available in ordinary telephones.

Tones are detected at the subscriber interface of the telephone exchange and, if necessary, signaled further to the other exchanges through which the connection is to be established. All digital local exchanges have the capability to use either pulse- or tone-dialing on a subscriber loop. The subscriber is able to select with a switch on his/her telephone which one they will use. Tone dialing should always be selected if the local exchange is a modern digital one.

Advantages of the tone dialing are:

- Quicker: dialing all digits takes the same time;
- Less dialing errors;
- Operation of end-to-end signaling;
- Additional push buttons *, #, (A, B, C, D) for activation of so-called supplementary services.

Supplementary services enable subscribers to influence the routing of their telephone calls. These services, such as call transfer, are not available with telephones that use pulse dialing. To control these services we need control buttons * and # that are available only on push-button telephones that use tone dialing.

We also use tone dialing to control *value-added services*. Value-added services are services that we can use via the telephone network but are usually provided by another service provider, not the telecommunications network operator. One example of value added services is telebanking. Tones are transmitted on the same frequency band as voice, and during a call we are able to dial digits to tell, for example, our discount number and security codes to the telebanking machine.

The worst disadvantage of a subscriber telephone is still a poor man-machine interface that makes new services difficult to use. Some telephones that have displays are more user-friendly, but subscribers still have to memorize command sequences in order to use the new services of a modern telephone network.

2.4 Local-Loop and 2W/4W Circuits

Any use of telephone channels involves two unidirectional paths, one for transmission and one for reception. The local loop, which connects a telephone to a local switch is a two-wire 2W circuit that carries the signals in both transmission directions; see Figure 2.6. Even the *integrated services digital network* (ISDN) basic rate subscriber connection uses this same 2W local loop.

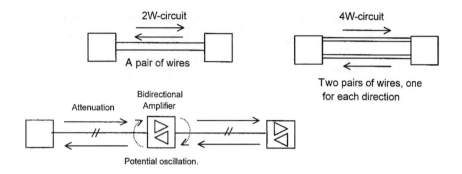

Figure 2.6 2W/4W connections. If the signal from one direction of a two-wire circuit is attenuated and needs to be amplified, there is a risk of oscillation.

Subscriber loops are and will remain two-wire circuits because they are one of the biggest investments of the fixed telephone network.

Early telephone connections through the network were two-wire circuits. Longer connections attenuate the speech signal, and amplifiers are needed on the line. In two-wire circuits amplification of a signal may cause oscillation or ringing if the output signal of an amplifier loops back to the input circuit of another transmission direction, see Figure 2.6.

The operation principle of electronics in the network is unidirectional, and inside the network we use two wires for each direction, 4W connections. Four-wire connections are also much easier to maintain because transmission directions are independent of each other and potential oscillation, as shown in Figure 2.6, is avoided. To connect a two-wire local loop to a four-wire network a circuit called a *2/4 wire hybrid* is needed.

We explain the operation principle of 2W/4W hybrid with the help of transformers that are easy to understand. A transformer consists of coils of wires around the same iron object. When an alternating current flows through one coil it produces a magnetic field into the iron core. This magnetic field generates current to the wires of other coils.

Figure 2.7 shows the 2W/4W hybrid in a subscriber interface of the telephone exchange. Two separate transformers are needed in the hybrid and both of them consist of three similar tightly coupled windings. In each transformer an alternating current in one coil generates an alternating current to all the other coils of the same transformer. Spots of coils indicate the direction of the current flow (polarity of the coil). In Figure 2.7 we see that the current of the receive pair generates two currents with opposite polarity through the other two coils of the transformer T2. These currents have opposite directions in transformer T1; they, or actually their magnetic fields in the iron core,

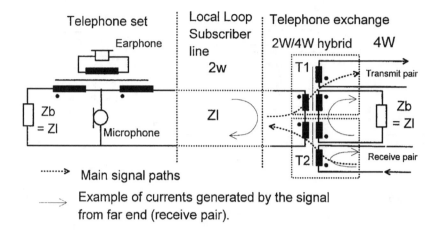

Main signal paths

Example of currents generated by the signal
from far end (receive pair).

Figure 2.7 Local-loop connection and 2W/4W hybrid. Dashed lines indicate the main
signal paths and the solid lines indicate the example of currents
generated by the signal from the far end (receive pair). Black spots indicate
the direction of the transformer wiring. Note that the signal current generates
a current with the same polarity through other wirings inside the same
transformer.

cancel each other; and the signal from the receive pair is not connected to the
transmit pair, or at least it is attenuated. In practice, the balance is not ideal
and the attenuated signal is connected back, which is heard as an echo from
the far end of the telephone circuit if the two-way propagation delay of the
circuit is long enough.

The satellite connections have long propagation delay and connections
between the digital cellular and fixed telephone networks have long coding
delay that causes a disturbing echo. Hence, in the case of these connections,
we have to use special equipment known as *echo chancellers* in the network in
order to eliminate the disturbing echo.

The 2W/4W hybrid performs the following operations:

- Separates the transmit and receive signals;
- Matches the impedance of the two-ire local loop to the network circuit;
- Provides a loss to signals arriving on the receive path, preventing them
  from entering the transmit path, which would otherwise cause an echo.

ISDN basic rate interface has bidirectional 160-kbps data transmission
on a two-wire circuit (ordinary subscriber loop). There the transmission direc-

tions are separated with the help of digital signal processing, so-called echo canceling technology. Many applications use the transformer circuit described previously together with digital signal processing technology that improves performance.

In every subscriber set the same principle as the 2W/4W hybrid is used to attenuate the subscriber's own voice from the microphone to earphone; see Figure 2.7. The reader may imagine what happens when the microphone generates an alternating current in the telephone set of the figure.

2.5 Telephone Numbering

International telephone connection from any telephone to any other telephone is made possible by the unique identification of each subscriber socket in the world. In mobile telephone networks each telephone set (or subscriber card) has a unique number.

The numbering is hierarchical, and it has an internationally standardized country code at the highest level. This makes national numbering schemes independent from each other. The structure of a telephone number is presented in Figure 2.8.

In the following subsection we explain the fields of the telephone number that are shown in Figure 2.8.

2.5.1 International Prefix

An international prefix or international access number is used for international calls. It tells the network that the connection is to be routed via an international telephone exchange to another country. The international prefix may differ

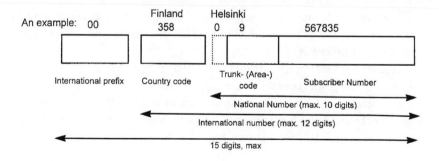

Figure 2.8 The structure of the telephone number. This CCITT-approved numbering scheme is the basis of international connections. The same principle is used in international data communication (e.g., X.25) networks.

from country to country, but it is gradually becoming harmonized. For example, all of Europe uses 00, but elsewhere it is still different. If there are many operators providing international telephone calls a subscriber may select different operators by using another prefix instead of 00, for example, in Finland 990 for Sonera, 994 for Telia Oy, and 999 for Oy Finnet International.

2.5.2 Country Code

The country code has two or three numbers that define the country of the B-subscriber. Country codes are not needed for national calls, because its purpose is to make the subscriber identification unique in the world. A telephone number including the country code is called an international number, and it has a maximum length of 12 digits.

Since there are a few hundred countries in the world, very many country codes are defined and their length varies from a single digit to four digits (some small areas have even longer codes). Examples of country codes are 1 for the United States and Canada, 49 for Germany, 44 for the United Kingdom, 52 for Mexico, and 1809 for Jamaica.

2.5.3 Trunk Code, Trunk Prefix, or Area Code

The trunk code defines the area inside the country where the call is to be routed. The first digit is a long-distance call identification and other numbers identify the area. The first digit is not needed in the case of an international call because that kind of calls are always routed via "the long-distance level" of the destination network.

In the case of cellular service the trunk code is used to identify the home network of the subscriber instead of the location. With the help of this network code a call is routed to the home network, which then finds out the location of the subscriber and routes the call to the destination.

The trunk code and the subscriber number together are a unique identification of a subscriber at the national level. This is called the national number, and its maximum length is 10 digits.

Trunk codes start with a 0 in Europe that is not dialed in calls coming from abroad. In some countries, the subscriber may select a long-distance network operator by dialing an operator prefix in front of the trunk code. For example, in Finland the long-distance operator numbers are 101 for Sonera, 1041 for Telia, and 109 for Finnet Oy.

2.5.4 Subscriber Number

The subscriber number is a unique identification of the subscriber inside a geographical area. For a connection to a certain subscriber the same number is dialed anywhere in the area. Because of the numbering hierarchy, the subscriber part of the telephone number of one subscriber may be the same as that of another subscriber in the another area.

2.5.5 Operator Numbers

In addition to the numbers mentioned previously, there may be a need to dial additional digits to select the service supplier (network operator), for example, to select a long-distance network or international connections; see prior explanation of international prefix and trunk code.

2.6 Switching and Signaling

In order to build up the requested connection from one subscriber to another, the network has switching equipment that selects the required connection. These switching systems are called exchanges. The subscriber identifies the required connection with signaling information that is transmitted over the subscriber line. In the network, signaling is needed to transmit the control information of a specific call and circuits between the exchanges.

2.6.1 Telephone Exchange

The main task of the telephone or ISDN exchange is to build up a physical connection between subscriber A, the one who initiates the call, and subscriber B according to signaling information dialed by subscriber A. The speech channel is connected from the time when the circuit was established to the time when the call is cleared. This principle is called the *circuit switching* concept, which differs from *packet switching,* which is used in data networks.

In the past the switching matrix was electromechanical and directly controlled by pulses from a telephone. Later on the control functions were integrated into a common control unit. Presently the common control unit is an efficient and reliable computer including huge real-time software. This kind of exchange is called *stored program control* (SPC) exchange; see Figure 2.9.

Every exchange between subscribers A and B connects a speech circuit according to signaling information that it is received from a subscriber or from the previous exchange. If the exchange is not the local exchange of subscriber B, it transmits signaling information to the next exchange, which then connects the circuit further.

2.6.2 Signaling

The control unit of the exchange receives the subscriber signaling, as dialed digits, from the subscriber line and makes consequent actions according to its

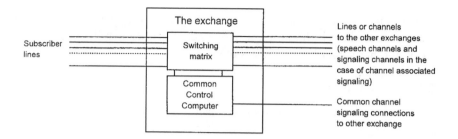

Channel Associated Signaling:
Each active telephone call has
own signaling channel between
exchanges in addition to speech
channel.

Common Channel Signaling:
One data channel between exchanges
is used for signaling purposes of
all telephone calls. This principle is
similar to computer communications
where messages are transmitted between
computers.

Figure 2.9 SPC exchange and signaling principles between exchanges.

program. Usually the call is routed via many exchanges and the signaling information needs to be transmitted from one exchange to another. This can be done via *channel associated signaling* (CAS) or *common channel signaling* (CCS) methods; see Figure 2.9.

2.6.2.1   Channel Associated Signaling

When a call is connected from a local exchange to the next exchange, a speech channel is reserved between exchanges for this call. At the same time another channel is reserved only for signaling purposes and each speech path has its own dedicated signaling channel while the call is connected. This channel can be, for example, a signaling channel in time slot 16 of a primary PCM frame as explained in the transmission section in Chapter 4. The main phases of signaling between exchanges are shown in Figure 2.10. The first speech channel and the related signaling channel are seized from exchange A to exchange B. Then the telephone number of B is transmitted to exchange B, which sends the ringing signal. When B answers, the speech connection is switched on and the conversation may start.

If subscriber B hangs up first, a *clear-back* (CBK) signal is transmitted from exchange B to A. Exchange A responds with a *clear-forward* (CLF) signal when subscriber A hangs up or when the time constant expires. The call is then disconnected by both exchanges.

There are many different signaling systems used for CAS, and some of them include additional signals that are not present in Figure 2.10. Signals

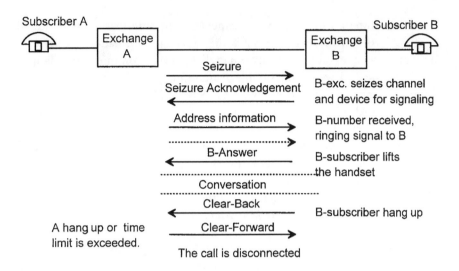

Figure 2.10 Channel associated signaling between exchanges.

that carry signaling information indicated in Figure 2.10 depend on the signaling system in use and may be, for example:

- Breaks of the loop between exchanges (loop/disconnect signaling);
- Tones with multiple frequencies, *multifrequency code* (MFC);
- Bit combinations of the signaling channel of a PCM frame.

CAS is still used in telephone networks, but it is gradually being replaced with a more efficient standardized method known as CCS.

2.6.2.2 Common Channel Signaling

The modern interexchange signaling system is called CCS. It is based on the principles of computer communications where information frames are exchanged between computers only when required. Frames include information about the connection to which the message belongs, address of the destination exchange, dialed digits, and information if the B-subscriber has answered. In most cases only one data channel between two exchanges is required. This is often one 64-kbit/s time slot of a primary 2- or 1.5-Mbit/s PCM frame, as explained in Chapter 4, and one channel is usually enough for all call-control communication between exchanges.

A widely used international standard of CCS is called CCS7, also known as *signaling system number 7* (SS7), CCITT#7, or ITU-T 7; and it is used in all modern telecommunications networks such as ISDN and GSM.

Establishing a call requires the same signaling information as indicated in Figure 2.10, but in the case of CCS the signaling information is carried in data frames that are transferred between exchanges via a common data channel.

In Figure 2.11 we see an example of an ordinary fixed network subscriber, subscriber A, calling subscriber B when CCS is used between exchanges in the network. The dialed digits are transmitted from subscriber A to the local exchange, as explained in Section 2.3. When a set of digits is received by exchange A, it analyses the dialed digits to determine to which direction it should route the call. From this information it looks up an address of the exchange to which it should send the signaling message for the call connection. Then the exchange builds up a data packet that contains the address of exchange B. This signaling message called *initial address message* (IAM) is then sent to exchange B. The remaining digits that did not fit into the IAM are then transmitted into one or more *subsequent address messages* (SAMs).

When all the digits that identify subscriber B are received by exchange B, it acknowledges this with an *address complete message* (ACM), to confirm that all the digits have been successfully received. This message also contains information if the call is to be charged or not and if the subscriber is free or not. Exchange B transmits the ringing tone to subscriber A and the ringing signal to subscriber B, and telephone B rings.

When subscriber B lifts the handset, an *answer signal charge* (ANC) is sent to start charging. Exchange B switches off the ringing signal and ringing

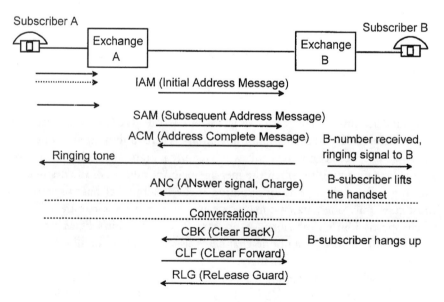

Figure 2.11 Common channel signaling between exchanges.

tone. Then both exchanges connect the speech channel and the conversation may start. When subscriber B hangs up, exchange B detects the on-hook condition and sends a CBK signal to exchange B. Exchange A responses with CLF signal. All exchanges on the line transmit the CLF message to the next exchange and each receiving exchange acknowledges it by a *release guard* (RLG) signal. This RLG message indicates to the receiving exchange that the connection is cleared and the channel is released and available for new calls.

2.6.3 Switching Hierarchy

During the early years of the telephone, the switching office or exchange was located at a central point in a service area and provided switched connections for all subscribers in that area. Hence, switching offices are still often referred to as central offices.

As telephone density grew and subscribers desired longer distance connections, it became necessary to interconnect the individual service areas with trunks between the central offices. With further traffic growth, new switches were needed to interconnect central offices and a second level of switching, trunk or transit exchanges, evolved. National networks typically have about five switching levels.

The actual implementation of the hierarchy and names of switching levels differ from country to country. Figure 2.12 shows an example of a possible network hierarchy [1] that is used in the United States.

The hierarchical structure of the network helps operators manage the network and makes the basic principle of telephone call routing straightforward;

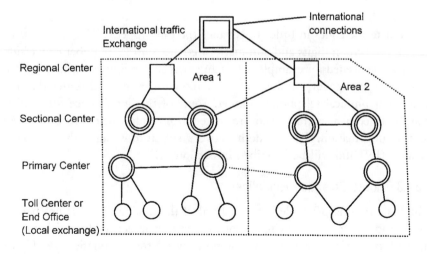

Figure 2.12 An example of switching hierarchy.

the call is routed up in the hierarchy by each exchange if the destination subscriber is not located below this exchange. As we saw, the structure of the telephone number supports this simple basic principle of routing in the switching hierarchy.

2.6.4 Telephone Call Routing

Calls that are carried by the network are routed according to a set of rules. The routing plan includes the numbering plan and network configuration.

2.6.4.1 Numbering Plan

The global rules for the highest level numbering, country codes and rules for overall numbering (e.g., maximum length), are given by CCITT, which is presently called ITU-T.

The numbering plan at the national level defines trunk or area codes in the country as well as certain nationwide service numbers (e.g., emergency numbers). These service numbers are defined to be the same wherever the call is originated. This national numbering plan is coordinated by the national telecommunications authority.

At the regional level, the numbering plan includes digits allocated to certain switching offices, exchanges, and the subscriber numbers for subscribers that are connected to a certain switch.

2.6.4.2 Switching Functionality for Routing

From the received code, a switching system must be able to interpret the address information, determine the route to or toward the destination, and manipulate the codes in order to advance the call properly. This includes the deletion of certain digits and automatic alternate routing. Number conversion may also be needed for example, when an emergency call dialed with a nationwide short emergency number has to be routed to a regional center that has a different physical telephone number. Some of this intelligence for routing may be stored in a centralized control system from which the exchanges request routing information. This modern network structure is called the *intelligent network* (IN) and will be described in Section 2.10.

2.6.4.3 Route Selection Guidelines

The basic routing principle is hierarchical: if the destination does not belong to the subscribers of the switch or of the switches under it, the call is routed upward; otherwise, it is routed to the port toward that destination; see Figure 2.13.

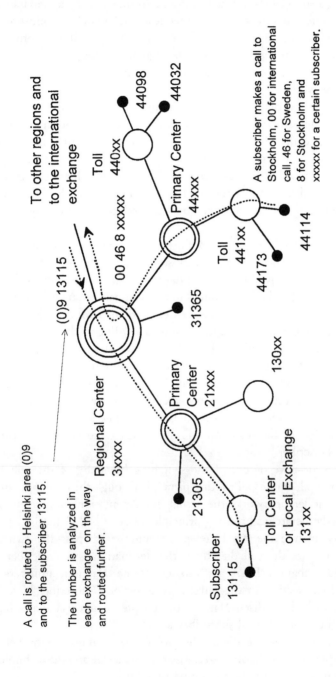

A call is routed to Helsinki area (0)9 and to the subscriber 13115.

The number is analyzed in each exchange on the way and routed further.

To other regions and to the international exchange

A subscriber makes a call to Stockholm, 00 for international call, 46 for Sweden, 8 for Stockholm and xxxxx for a certain subscriber.

(0)9 13115

00 46 8 xxxxx

Toll 440xx

44098

44032

Primary Center 44xxx

Toll 441xx

44173

44114

31365

Regional Center 3xxxx

Primary Center 21xxx

130xx

21305

Subscriber 13115

Toll Center or Local Exchange 131xx

Figure 2.13 Telephone call routing.

However, modern exchanges can do more than strictly hierarchical rout-ing. If there is a sufficient volume of traffic, calls may pass by a hierarchy level or may be connected directly to another low-level switch, as illustrated in Figure 2.12. This may be reasonable, for example, if the switching offices and their subscribers are on opposite sides of the regional border. The telecommuni-cation operator is free to define the detailed actual routing to optimize the usage of the network.

In the example of Figure 2.13, a Finnish subscriber makes a call to Stockholm, Sweden and dials the international prefix "00", country code "46" for Sweden, area code "(0)8" (leaves out zero) for Stockholm, and subscriber number "xxxxx". The international prefix is actually all that exchanges in Finland need to know. When exchanges in the switching hierarchy detect it, they route this call up toward the international exchange. The international exchange then analyzes the country code and selects an outgoing route to Sweden.

Another example in Figure 2.13 illustrates routing of a long-distance call to a subscriber from another region. A subscriber in another region dialed "09 13115" for a call to Helsinki. The first digit "0" tells the exchanges that this is a long-distance call and is to be routed to the regional exchange. The regional center is connected to other regional centers and then routes this call, with the help of other regional centers, to Helsinki according to the next digit "9". The regional center of Helsinki analyzes the next two numbers "13" and selects the route down to the primary center where these subscribers are located. The primary center then checks the numbers "131" and notices that this is not "my subscriber" but the subscriber is located "below me" and routes the call to the corresponding toll center or local exchange. The local exchange selects the subscriber loop of the telephone number 13115 and connects a ringing signal to the subscriber.

In this section we described the switching hierarchy of the telephone network and the telephone call routing principle through the exchanges in this hierarchy. In modern networks the actual implementation may be different from this strictly hierarchical routing principle we described. A local telephone exchange may analyze the whole telephone number, bypass the switching hierarchy, and route the call directly if the destination is a subscriber of a neighbor local exchange. There are also sets of the telephone numbers that have no fixed connection to the physical location of a subscriber loop. The IN technology, which we discuss later in this chapter, connects a dialed logical number and a certain physical subscriber loop.

In the next section we divide the global telecommunications network into three simplified layers in order to clarify their structure and the technologies that are used to implement their required functions.

2.7 The Local Access Network

The local access network provides the connection between the customer's telephone and the local exchange. Ordinary telephone and ISDN subscribers use two wires, a pair, as a subscriber loop, but for business customers a higher capacity optical fiber or microwave radio may be required. Many different technologies are used in a local access network to connect subscribers to the public telecommunications network. Figure 2.14 illustrates the structure of the local access network and shows the most important technologies in use.

Most subscriber connections use twisted pairs of copper wires. Subscriber cables contain many pairs that are shielded with common aluminum foil and plastic shield. In urban areas cables are dug into the ground and may be very large, having hundreds of pairs. Distribution points that are installed in outdoor or indoor cabinets are needed to divide large cables into smaller ones and distribute subscriber pairs to houses as shown in Figure 2.14. In suburban or country areas overhead cables are usually a more economical solution than underground cables.

Optical connection is used when a high transmission capacity (more than 1.5 Mbit/s to 2 Mbit/s) and/or very good transmission quality is required. Microwave radio is more economical than optical fiber when there is a need to increase data capacity beyond the capacity of an existing cable network.

A new technology for implementation of ordinary subscriber loops for fixed telephone service is WLL or RLL. It uses radio waves and does not require the installation of subscriber cables; it is a quick and low-cost way to connect a new subscriber to the public network. With the help of this technology, new operators are able to provide services in an area where another operator owns the cables.

When there is a need to increase cable network capacity for subscriber connections, it may be more economical to install *concentrators, remote subscriber units,* or *subscriber multiplexers* in order to utilize existing cables more efficiently. We use one of these three terms to describe the switching capability of the remote unit. Concentrators may be capable of independently switching local calls between the subscribers connected to it. A remote subscriber unit is basically the subscriber interface part of the exchange that is moved close to the subscribers. Subscriber multiplexers may only connect each subscriber to a time slot (channel) in the PCM frame. The detailed functionality of these systems depends on the manufacturer, but we may say that they connect only those subscribers to the local exchange who have picked up their handsets. Digital transmission between an exchange and a concentrator further improves utilization so that two cable pairs may serve tens of subscribers.

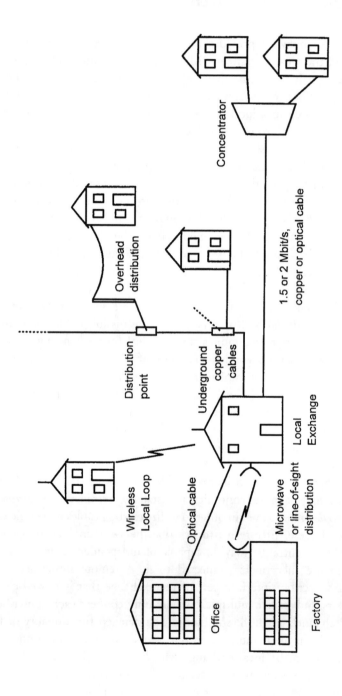

Figure 2.14 An example of a local access network.

2.7.1 Local Exchange

Local or subscriber loops connect subscribers to local exchanges, which are the lowest level exchanges in the switching hierarchy. The main tasks of the digital local exchanges are to:

- Detect off-hook condition, analyze the dialed number, determine if a route is available, and if the called customer is free;
- Connect the customer to another in the same local area;
- Connect the customer to a trunk exchange for longer distance calls;
- Provide metering and collect charging data for its own subscribers;
- Convert two-wire local access to a four-wire circuit of the network;
- Convert analog speech into a digital signal (PCM).

The size of local exchanges varies from hundreds of subscribers to tens of thousands subscribers or even more. A small local exchange is sometimes known as a remote switching unit (RSU), and it performs the switching and concentration functions just as all local exchanges do. Local exchange reduces the capacity (number of speech channels) typically by a factor of ten or more, that is, the number of subscribers of the local exchange is ten times higher than the number of trunk channels from the exchange for external calls. Figure 2.15 shows some different subscriber connections to a local exchange and how they are physically installed.

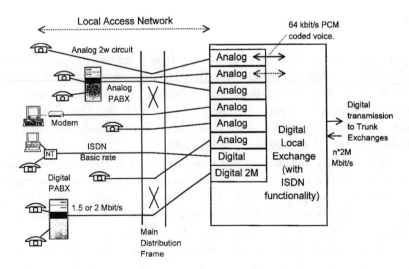

Figure 2.15 Local access network and digital local exchange.

All subscriber lines are wired to the *main distribution frame* (MDF), as shown in Figure 2.15, which is located close to the local exchange. It is a large construction with a huge number of connectors. Subscriber pairs are connected to one side and pairs from the local exchange to the other. Between these connector fields there is enough space for free crossconnection. Cables and connectors are usually arranged in a logical way considering the subscriber cable network structure and switching arrangements. This fixed cabling stays the same over long periods of time, but connections between sides change daily, for example, because a subscriber moves to another house in the same switching area.

A crossconnection in the MDF is usually made with twisted open pairs that are capable of carrying data rates up to 2 Mbit/s. Ordinary subscriber pairs are used for analog telephone subscribers, analog PBX/PABX connections, ISDN basic rate connections, and digital PBX/PABX connections.

A digital exchange may include both analog and digital subscriber interfaces, and 1.5- or 2-Mbit/s digital interfaces are available for *private (automatic) branch exchange* (PBX/PABX) applications.

If the local switch has ISDN-capability, basic rate and primary rate interfaces are available. Ordinary subscriber pairs are used for ISDN basic rate connections (160-kbit/s bidirectional) and an NT is required in customer premises. The primary rate interface of ISDN (1.5 Mbit/s or 2 Mbit/s) is used for PABX-connections and requires two pairs, one for each transmission direction.

2.8 The Trunk Network

As we saw in Section 2.6, the national switching hierarchy includes multiple levels of switches above local exchanges. Figure 2.16 shows a simplified structure of the network where levels above local exchanges are shown as a single level of trunk exchanges. The local exchanges are connected to these trunk exchanges, which are linked together to provide a network of connections from any customer to any other subscriber in the country.

High-capacity transmission paths, usually optical line systems, with capacity up to 2.4 Gbit/s, interconnect trunk exchanges. Note that there are alternative routes in a transport network. If one of these transmission systems fails, switches are able to route new calls via other transmission systems and trunk exchanges to bypass the failed system; see Figure 2.16.

All transmission systems that interconnect trunk exchanges are called a transmission or transport network. Its basic purpose is simply to provide a required number of channels (capacity) from one exchange site to another.

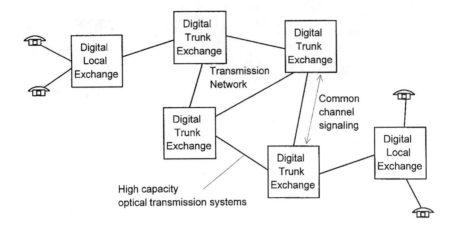

Figure 2.16 Links between trunk and local exchanges.

Exchanges use these channels of the transport network for calls that they route from one exchange to another on subscriber demand.

The trunk exchanges are usually located in major cities. They are digital and use the international CCS standard SS7 to exchange routing and other information between exchanges.

2.9 The International Network

Each country has at least one international switching center to which trunk exchanges are connected, as shown in Figure 2.17. Via this highest switching hierarchy level international calls are connected from one country to another and any subscriber is able to access any of the other 800 million subscribers around the world.

High-capacity optical systems interconnect international exchanges or switching centers of national networks. Submarine cables (coaxial cable or optical cable systems), microwave radio systems, and satellites connect continental networks to make up the worldwide telecommunications network.

The first submarine cable telephone system across the North Atlantic Ocean was installed in 1956, and it had the capacity of 36 speech channels. Modern optical submarine systems have a capacity of several hundred thousand speech channels, and new high-capacity submarine systems are put into use every year. These are the main paths for intercontinental calls. Satellite systems are used as back-up channels in the case of congestion.

We described the common structure of the global telecommunications network without a separation of different network technologies. We need

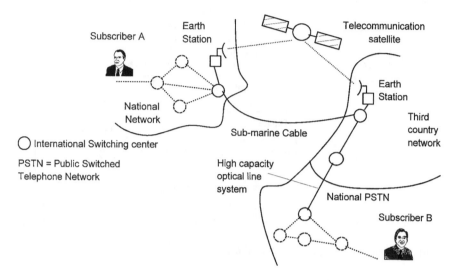

Figure 2.17 The international network. A call from subscriber A to subscriber B requires a so-called transit connection through a third country network. A satellite connection is used if all the channels of the submarine cable system are reserved.

different network technologies to provide different types of services; and the telecommunications network is actually a set of networks, each of them having characteristics suitable for the service it provides. In the next section we describe briefly the most important network technologies. We will describe most of these in more detail in later chapters.

2.10 Telecommunications Networks

To this point we explained the operation of a *public-switched telecommunications network* (PSTN) and cited the conventional telephone networks as an example. However, PSTN contains many other networks that are optimized to provide services with different characteristics. We review these different network technologies in this section.

There are many ways to divide telecommunications networks into categories. If we consider the customers of networks and the availability of services, there are two broad categories: public networks and private or dedicated networks.

2.10.1 Public Networks

Public networks are owned and managed by telecommunication network operators. These network operators have a license to provide telecommunication services and that is usually their core business. Any customer can be connected to the public telecommunications network if they have the correct equipment and an agreement with the network operator. All the switched (dial-up) public networks that we describe in the following subsections are referred to as PSTN. The same acronym PSTN is often used to refer to *public-switched telephone network* only.

2.10.1.1 Telephone Network

The telephone network is the main public network in use and is sometimes referred to as *plain old telephone service* (POTS). It is used to provide voice communications, although data can be substituted for speech with the help of a modem. We described the operation of this network in the previous sections of this chapter.

2.10.1.2 Mobile Telephone Networks

Mobile telephone systems provide radio communication over the local access part of the network. They are regional or national access networks and connected to the PSTN for long-distance and international connections. We will describe mobile networks in Chapter 5.

2.10.1.3 Telex Network

The telex network is a telegraph network that allows teleprinters to be connected via special dedicated switches. The bit rate of telex is very slow, 50 or 75 bits per second, which makes it robust. It was once widely used, but its importance is reduced since other messaging systems like electronic mail and fax have reduced its share.

2.10.1.4 Paging Networks

Paging networks are unidirectional only. Pagers are low-cost, light-weight wireless communication systems for contacting customers without the use of voice. Simple pagers just say "beep," but more sophisticated pagers can receive large amounts of text and display the message on a screen.

2.10.1.5 Public Data Networks

Public data networks provide leased point-to-point connections or circuit-switched or packet-switched connections. *Leased point-to-point* line is often an economical solution for connections between the LANs of corporate offices in a region. Circuit-switched networks dedicated to data transmission are not widely used today. Kilostream in the United Kingdom is one example of a circuit-switched data service. Packet-switched data service is provided by the X.25 network worldwide. It operates according to the X-series recommendation of ITU-T, but marketing names of X.25 networks are different from country to country, for example, Auspak in Australia and Finpak in Finland.

2.10.1.6 Internet

The Internet is a worldwide packet switched network developed from the ARPANET that in turn was developed in the late-1960s by the U.S. Department of Defense. The ARPANET was developed into a wide area computer network, the Internet, which was mostly used in the 1970s and 1980s by academic institutions. In the first half of the nineties the user-friendly graphical user interface World Wide Web was introduced; since then, the use of the Internet has expanded very rapidly. Presently the Internet is the major information network in the world and there are many *Internet service providers* (ISPs) that provide Internet services for both businesses and residential customers. The rapid expansion of the Internet is expected to continue. The evolving commercial services (e.g., electronic shopping), the new access technologies (such as xDSL; see Chapter 4), and speech and video service will further increase its importance in the near future.

2.10.1.7 Integrated Services Digital Network

The present telephone network will gradually be developed into ISDN, in which all information is transmitted in digital form from end to end. With the help of some hardware and software updating, most modern digital telephone exchanges will be able to provide ISDN service. The main modification required, in addition to the replacement of signaling and other software, is the replacement of analog subscriber interface units with digital ones, as shown in Figure 2.18.

The ordinary two-wire subscriber loop of the telephone network is upgraded to the basic rate access of ISDN by a *network terminal* (NT) on the subscriber premises and by a basic rate interface unit and ISDN software in the local exchange. The bidirectional data rate in the subscriber loop is 160 kbps, which carries 144 kbps of user data and additional framing information. With the help of framing information the receiving end is able to distinguish different channels from the data stream. User data contains two independent 64-kbps circuit-switched user channels, B-channels, and a 16-kbps signaling channel, D-channel. Subscribers may use user channels, B-channels at 64 kbps, for ordinary speech transmission, data, facsimile, or video conferencing connections.

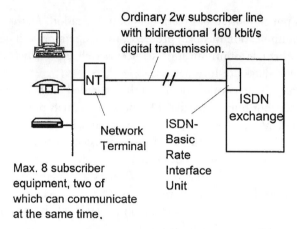

Figure 2.18 Public networks and ISDN. Basic rate, 2B + D, subscriber access to ISDN. Routing B-channels, 64 kbit/s, is independent. D-channel, 16 kbit/s, is used for signaling. The total information rate is 144 kbit/s, which makes 160 kbit/s when framing information is added.

Subscribers may use both B-channels independently at the same time and dial them up independently, for example, using one of these channels for telephone and another for Internet connection. ISDN provides a full 64-kbps connection end-to-end, which is more than double the data rate available for subscribers using an analog subscriber pair and modem transmission over the telephone network circuit.

Users may connect up to eight terminals to a NT and two of them may be in use at the same time. The advantages of ISDN over the analog telephone service are a higher data rate and the availability of two connections at the same time. This technology has been available for some time, but its usage has been low because of high tariffs. Today operators offer attractive tariffs, and the increased demand of high-performance Internet connections is making it more and more popular. The existing low-cost technology makes it feasible for network operators to provide ISDN connections at a lower cost than two ordinary analog telephone connections in some cases.

2.10.1.8   Radio and TV Networks

Radio and television networks are usually unidirectional radio distribution networks for mass communications. Traditionally the operators of these networks have not provided dial-up bidirectional telecommunication services. Access to these networks is presently available in urban areas via cable-TV networks built by cable-TV operators. These operators have not been allowed to provide other telecommunication services, and their wideband cable network to homes has not supported bidirectional communication. As the deregulation of the telecommunication business proceeds, these operators will become active in providing other telecommunication services as well, especially fixed telephone service and high data-rate Internet access.

The cable-TV networks are presently upgraded with technologies that allow subscribers not only to receive TV and radio signals but to transmit data to the network. Most of the investment was already made when wideband cables were installed. This existing medium is especially attractive for providing Internet service to every home connected to a cable-TV network. Low-cost and simple-access terminals that use a TV-set as a display and a TV remote control as a control device will provide user-friendly access to Internet for everyone. This development will increase importance of the present cable-TV operators in telecommunication business.

2.10.2 Private or Dedicated Networks

Private networks are built and designed to serve the needs of particular organizations. They usually own and maintain the networks themselves. Services provided are a tailored mix of voice, data, and, for example, special control information.

2.10.2.1 Voice Communication Networks

Examples of private dedicated voice networks are the police and other emergency services and taxi organizations using *private mobile radio* (PMR). Railway companies also have private voice networks using cables that run alongside the tracks.

2.10.2.2 Data Communication Networks

Data communication networks are dedicated networks that are especially designed for the transmission of data between the offices of an organization. They can incorporate LANs with main frame computers feeding information to the branch offices. Banks, hotel chains, travel agencies etc. have their own separate data networks to update and distribute credit and reservation information.

2.10.3 Virtual Private Networks

It is very expensive for an organization to set up its own private network. Another choice is to lease resources, which are also shared with other users, from a public network operator. This *virtual private network* (VPN) provides a service similar to an ordinary private network, but the systems in the network are the property of the network operator.

In effect, a VPN provides a dedicated network for the customer with the help of public network equipment. As companies concentrate more and more on their core businesses, they are willing to outsource the provision, management, and maintenance of their telecommunication services to a public network operator that has skilled professionals dedicated to telecommunications.

The principle of VPN is used for voice services such as corporate PBX/PABX networks. In this case the network that interconnects the offices of a company uses channels of the PSTN that are leased from a public network operator.

A new important application of VPN is *intranet*. It means a private data network that uses open Internet technology. Physically an intranet may contain many LANs at different sites. For their interconnections a VPN is established to provide data transmission between sites through the public Internet network. Note that Internet uses the packet-switching principle and there are no physically separate channels for each VPN as in the previously explained voice VPN. Since the packets are not separated into dedicated point-to-point channels, there is a security risk when the public Internet is used for interconnections instead of leased lines or a circuit-switched network such as ISDN. To overcome this problem, *firewalls* are used in an intranet at the interface between each LAN and the public Internet. The firewalls perform the authentication of the communicating parties and encrypt data for transmission through the public Internet.

Another network related to intranet is *extranet*. It means connections between selected users of the Internet and an intranet. These external users of a private intranet may be, for example, customers or material suppliers. Extranet uses Internet technology as well, and for security reasons firewalls are used for user authentication and data encryption.

2.10.4 Intelligent Network

A conventional telephone network is able to establish a connection only to a socket that is identified by a number of a B-subscriber. There is no "intelligence" in this kind of operation, dialing a certain number makes a connection to a certain socket. Connection set-up is always done in the same way, whether the intended B-subscriber is available or not.

In the old days a human operator made the switching on a switchboard. If an operator knew that the called party was presently visiting his neighbor, she might connect the call directly to the neighbor's phone. There was some "intelligence" in the network that improved accessibility. In a modern telecommunications network this intelligence is implemented with help of IN technology.

The IN is an ordinary digital telephone network with some additional capabilities like flexible call routing and voice notifications. Traditionally a telephone number has been the identification of a certain physical subscriber line and a *socket*. In IN the physical number and service number have no fixed relation and it may also change with time. For example, emergency service may be available in the daytime in multiple locations but at nighttime only in one location of the area.

2.10.4.1 Distributed Intelligence

Network operators implement supplementary services, such as call forwarding, to assist subscribers in making successful calls. This increases the number of successful calls, the utilization of the network, and, as a consequence, operator's revenue from call fees. We may implement these services by updating corresponding functions to each local exchange. However, when more and more services have been introduced, updating new services becomes a great burden to the network operator. The structure of IN was developed to help network operators and service providers introduce, update, and develop new services in a more efficient way.

2.10.4.2 Centralized Intelligence

A certain range of telephone numbers is reserved for IN services only. The basic structure of an IN is illustrated in Figure 2.19. When a SSP, which

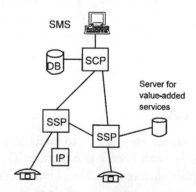

Figure 2.19 The structure of the intelligent network. SMS, service management center, for updating services or introducing new ones; SCP, service control point, which gives routing and charging information to switches; DB, data base stores the service information, for example, number conversion for a call transfer; SSP, service switching point, telephone exchange that requests routing information from SCP if the IN number is detected; IP, intelligent peripheral, gives the voice notifications if required.

performs the functions of an exchange, detects an intelligent network service number, it requests routing information from the SCP, which provides information about how this call should be handled.

2.10.4.3 The Structure of Intelligent Network

IN technology makes providing new services efficient with the help of control data that is centralized and available to all SSPs switches. Otherwise, service information should be updated to all exchanges when there is a need to change it.

Figure 2.19 shows the main network elements of IN. In principle, we could implement all intelligence in SCP and its data base could store all the routing information. This would require heavy signaling between the switching points and the SCP. In practice, the services that do not require a centralized data base are implemented in switching points to reduce the load on the SCP and the signaling connections between SCP and SSPs. A simple example is call transfer, which can be handled independently by local exchanges and thus does not require signaling between the SSP and SCP. This is called *distributed intelligence* that we use to implement most *supplementary services.*

Examples of supplementary services include the following:

- *Call forwarding* permits one to direct incoming calls to another telephone. Forwarded calls are regarded as being made from your home telephone and will therefore charge to the telephone bill of the subscriber that has made the call forward.

- *Call waiting* means that, during a call in progress, a subscriber is notified of an incoming call. You hear the message as a faint tone in the receiver, while the caller simultaneously hears a normal ringing tone. You can alternate between these two calls.

- *Automatic callback* can be used when the number you are trying to call is busy. A subscriber notifies the system that you want to have a call established when the called party becomes free and she/he will be informed when this happens. When the subscriber then lifts the receiver, the number will automatically be redialed.

- *Abbreviated dialing* permits a subscriber to specify short numbers that correspond to complete telephone numbers you use most frequently.

- *Screening of incoming and outgoing calls*: A subscriber may specify which telephone numbers are forbidden and calls to or from the specified numbers are not established. This service is implemented by the telephone service provider according to a customer request. A subscriber may, with the help of this service, avoid charges that may be very high when expensive service numbers are called from her/his telephone.

An important category of services implemented with help of IN technology is *value-added services*. This term refers to the services that give additional value, not just point-to-point telephone conversation. These services are often provided by separate service providers, not the telecommunication service provider.

Some examples of IN services include the following:

- *Universal access number*: A company with several offices in different parts of a country may have the same number throughout the country. Each call is automatically connected to the office closest to the calling subscriber. The cost of the call is the same no matter to which office the call is connected.

- *Premium rate services*: Information is provided over the phone, for example, a doctor or lawyer service: The service provider charges subscribers on their telephone bill. The charge is dependent on the called service number.

- *Freephone*: This service is used by companies that want to provide free customer service; the receiver pays for the call.

- *Credit-card call*: A service user can pay with his credit card by dialing his account number and identity code.

The modern telecommunications networks using IN technology provide many other services and a few new ones appear annually. An example of these is inexpensive home-to-mobile and mobile-to-home calling for which one dials a specific number given by an operator. This service gives a competitive advantage to an operator who provides both cellular and fixed service. Another example is a card service for which a serviceman dials a specific service number and security code and the network operator charges his employer instead of the telephone he is calling from.

A concept called *universal personal telecommunication* (UPT) will be implemented with IN technology. This service means that the telephone number is not associated with a certain telephone set or socket but the subscriber may receive their calls anywhere. Using a personal code, a subscriber identifies himself to the system and indicates the terminal to which his call shall be routed.

In previous sections we described the structure and operation of the telephone network and looked at different network technologies that we need to provide different services. In the following section we look at how all of this fits together.

2.10.5 Public-Switched Telecommunications Network Today

The overview of modern PSTN is presented in Figure 2.20. The structure and functionality of the network are only reviewed here because most of the elements in the figure are discussed in other sections of this book.

Figure 2.20 presents a simplified diagram of a regional or national PSTN that has connections to the Internet and the international telecommunications network. PSTN contains the *public land mobile network* (PLMN), which provides wireless access and is connected to the telephone/ISDN network at the trunk exchange level. PSTN also contains the X.25 *packet-switched public data network* (PSPDN), which is dedicated to providing data services. It is connected to the telephone network, so telephone network subscribers may access it with the help of a modem.

Some different means of accessing telecommunications networks are also shown in Figure 2.20. Digital PBX/PABX is connected to a local exchange with a 1544/2048-kbps digital line that has the capacity of 24/30 simultaneous calls. This connection is called primary rate interface in the case of ISDN. PBX/PABX is a dedicated small exchange that provides telephone service to the personnel of a company. Analog PBX/PABX uses analog telephone lines, one for each simultaneous external call. Each analog line (cable pair or two pairs) carries one telephone call with signaling. This analog signaling is close to the ordinary analog subscriber loop signaling that we described previously.

The corporatewide PBX/PABX service can also be implemented without any equipment investments in the company, that is, without physical PABX equipment. Network operators provide a service called *Centrex* and for that the public exchange is programmed to behave as a PBX/PABX. One of the subscriber lines is programmed to operate as a switch board line and the others make up a user group with abbreviated dialing and other PBX/PABX services.

For data communication via an analog network or a digital network with analog subscriber interfaces a modem is required. In ISDN service, which is fully digital, no modem is needed and an end-to-end high data-rate digital circuit is available with the help of a NT, which takes care of the digital bidirectional transmission over the subscriber loop.

The connection from an office to a PSPDN usually requires a leased line from the customer premises to the nearest packet-switched node that performs the switching function in a X.25 data network. Leased lines are often the most economical solution for the high data-rate circuits that are needed, for example,

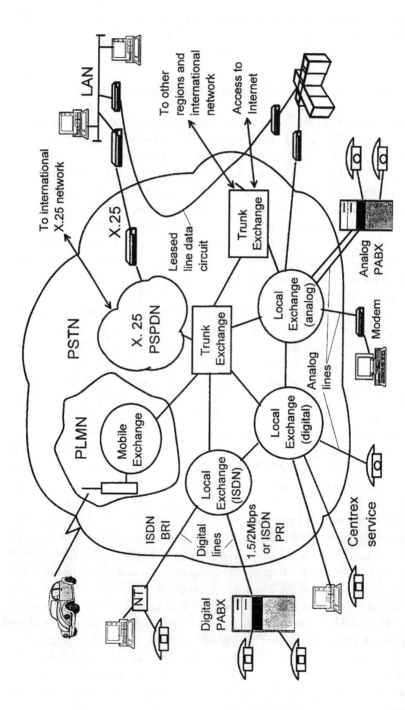

Figure 2.20 Public-switched telecommunication network today.

for LAN interconnections. Different options for data connections are discussed in Chapter 6.

As we have seen, telecommunications networks contain a huge number of different complex systems that are located in multiple sites. In the olden days, when the structure of the network was simple, most of the equipment sites had personnel to keep systems operational. This personnel carried out fault location and performed needed maintenance operations. Nowadays systems are so many and so complicated that this way of network operations and maintenance is no longer possible and implementation of automated network management tools is mandatory for all network operators. The following section gives an overview of the importance of network management and of the standardized structure of network management.

2.11 Network Management

The importance of network management has grown together with the size and complexity of the telecommunications network. The standardization of this area is not as advanced as the standardization of telecommunication systems that carry the actual traffic and provide the services. Efficient network management is a key tool in helping a network operator improve his services and make them more competitive.

2.11.1 Introduction to Network Management

Traditionally, systems that take care of control and supervisory functions in a telecommunications network have been known as *operation and maintenance* (O&M) systems. Nowadays we prefer to use the term *network management* because the functions performed by network management include much more than the conventional operation and maintenance systems.

Operation functions cover subscriber management functions and enable the network operator, for example, to collect charging data and move and terminate subscriptions. Operation also includes traffic monitoring and controlling the network in such a way that the risk of overload is minimized, for example, by switching traffic from overloaded connections to other systems.

Maintenance includes monitoring the network and, when a fault occurs, corrective actions are performed. Bit error rates and other parameters are continuously measured for the early detection of faults. When a fault is detected, the operator's staff starts troubleshooting in order to localize the fault. This has previously been quite a difficult task because it was done manually and many systems may detect a fault even when the actual fault may be in only one of them or even somewhere else. Maintenance, like other network management functions, is becoming more and more computerized, making fault location easier and quicker with the help of centralized management systems that provide graphical information about the network condition.

2.11.2 Who Manages Networks?

Corporate networks are private networks that contain LANs, which are interconnected by circuits provided by a public telecommunications network operator. We can divide corporate networks into two areas of network management responsibility: local networks in corporate sites and interconnections between sites implemented in a public network that provides interconnections as shown in Figure 2.21. The corporate networks are managed by the people responsible for network operation inside a company.

Management is often divided hierarchically. Local or site managers only take care of LAN networks at each office. A centralized organization of a company manages the usage availability of the *wide area network* (WAN) connections between sites. A centralized organization offers service to business units at various sites and optimizes the utilization of expensive long-distance or even international WAN connections.

The main concerns of network managers of a company include:

- Network change management (hardware updates);
- The location and repair of malfunctions;
- Software updates and version control;
- Network security.

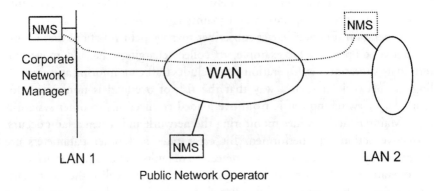

Figure 2.21 Management responsibility of a corporate network.

Most present network elements of LANs provide network management functions via a standardized management interface. This open standard is known as *Simple Network Management Protocol* (SNMP). Software packets for centralized management work stations of LANs are commercially available.

The public network operator manages the public network in order to be able to provide reliable service to customers. Network optimization to avoid unnecessary investments as well as quick repairs in the case of faults is important. Short delivery times of leased line circuits are an important competitive advantage today, and a network operator can make delivery time shorter with the help of sophisticated network management tools.

In addition to private network management needs, accounting functions are needed in a public network for switched circuits. For example, in the case of packet-switched service, the amount of transferred data must be recorded to generate bills to customers.

Public networks contain many different technologies, and the operator's organization is usually divided into different responsibility areas, such as transmission, telephone exchanges, leased-line data networks, and packet-switched data services. Today these organizations usually have their own dedicated and incompatible network management systems, probably with some kind of geographical hierarchy, and the integration of these is an important issue for the future. At least some level of integration is needed because all services usually use the same transmission network. To solve this problem, ITU-T has defined a common management concept known as the *telecommunications management network* (TMN). In the following section we describe the data communications network (DCN) as it pertains to the TMN concept and which is responsible for the transmission of management data.

2.11.3 Data Communications Network of TMN

Not only different networks but even network elements (equipment) may have their own operation and maintenance systems that may be incompatible today. As a consequence, if a fault occurs in the network, the network operator's personnel may have to use several different O&M systems for fault localization. ITU-T has worked several years to define the vendor-independent network management concept TMN.

In the TMN concept of ITU-T the transmission of management data between management work stations and network elements is separated from the transmission of user data as shown in Figure 2.22. The transportation network of management data is the DCN.

Even though DCN is a logically separate network from the actual telecommunications network, the management messages often use the same network

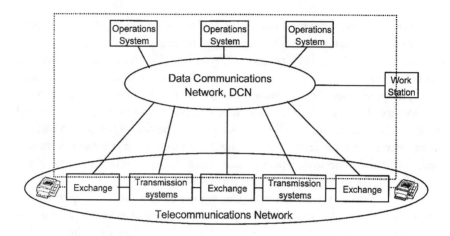

Figure 2.22 Data communications network.

as the actual telecommunication services. Most transmission systems, such as SDH as described in Chapter 4, provide data channels for network management purposes. This requires careful planning of the DCN because a fault on a transmission link may disturb management messages that are necessary for fault localization. Therefore, the design of the DCN must be as independent as possible from the network that transmits user data.

Sometimes a network operator can physically separate management data from user data using another network for management links. For example, X.25 may be used for telephone network management. The use of another network may also be feasible in order to implement redundant routes to DCN; that is, the management data is sent via another connection when the one in use fails.

2.11.4 Telecommunications Management Network

The overall management concept that ITU-T has defined is known as TMN. The standardization of TMN is aiming to cover all aspects in order to make the centralized operation and maintenance of telecommunications networks possible in a multivendor environment.

The complete standardization of TMN aims to cover the specifications of:

- The *physical architecture of TMN*: what systems are needed in TMN, and how they are interconnected;
- *Interface protocols*: how network elements and management systems exchange information (the structure and types of messages);

- *Management functions*: what functions in the network elements should the network management system be able to access;

- *Information model*: for each different system in the network, how each manageable function (in detail) is described in management messages.

The recommendation for TMN, including the definition of TMN as a concept, was approved by ITU-T (CCITT at that time) in the middle of the eighties. This defines the physical architecture of TMN, which is shown in Figure 2.22. TMN is understood to be separate from the actual telecommunications network, though network systems must provide the management interfaces and management functions that they are able to perform.

The physical architecture of TMN (see Figure 2.23) contains:

- *Operations system* (OS) for centralized network management;

- *Data communications network* (DCN) for management data transfer;

- *Mediation devices* (MD) to adapt proprietary management interfaces to standardized Q3 interface;

- Management functions integrated in the *network elements* (NE) of the telecommunications network.

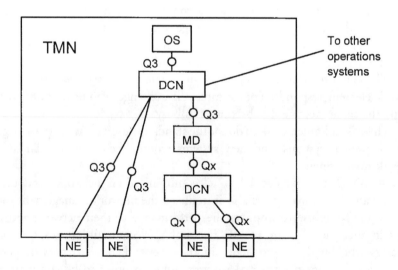

Figure 2.23 Physical architecture of telecommunications management network (OS, operating system; DCN, data communications network; MD, mediation device; and NE, network element).

The most important and most difficult standardization issue is the specification of management interfaces, Q3. Lower level protocols, like the physical network that carries actual data and formats of messages, are already standardized, but there is still much work to do in the area of information models. The specification work of information models is an endless task because new systems require their own models and an update of a system often requires a revision of the information model. The information model defines the managed objects (manageable resources) of a system and their relationships. The specification of an information model is mandatory before we talk about vendor-independent management.

The information model is specified by the *management information tree* (MIT), which defines all the managed objects in a system. The managed objects contain all the resources that the management system can access. Each managed object has a unique identification that consists of a sequence of names (or numbers) starting from the root and having multiple options at each level. For example, at the second level after the root we have one branch for the ISO (1) and another for the CCITT (0), and an object identifier contains ISO if this is specified as the right path to our system and its managed objects. The highest levels of the MIT are standardized, but the compatibility of the systems from different vendors requires detailed standardization down to the managed object and its behavior.

For example, if we want to get information about whether subscriber 1 of an exchange is busy, we must have a complete specification of what kind of message, when transmitted to the exchange, will produce the wanted response regardless of the manufacturer of that exchange. The lower level protocols define the structure of the messages, and the information model must specify in detail the information content of the management message with which the network element responds. For example, all exchanges should respond with exactly the same message if subscriber 1 is busy.

There is still much work to do for the standardization of network management of present systems, and new systems require their own standards for network management.

In this chapter we looked at telecommunications networks and their structure and functionality; we also introduced the network management that network operators use to improve the performance of their network and to maintain their network in an effective way. Telecommunications network operators who build up and maintain their network have to provide good service at as low an investment level as possible in order to be competitive. Their problem is how to minimize investment but still keep customers happy. To find out where they should invest and what the bottlenecks of the network are, they continuously perform traffic engineering, which is introduced in the next section.

2.12 Traffic Engineering

Traffic engineering is a key issue to telecommunication network operators in keeping customers (subscribers) happy while minimizing network investments. Nowadays network operators have to pay more and more attention to these aspects because of increasing competition in the telecommunication services market. The capacity of the network (i.e., number of channels between exchanges, exchange sizes, and number of radio channels in a cellular network) should be increased where the bottlenecks of the network are found. Therefore, the utilization of the network is continuously measured and traffic demand in the future estimated.

An important capacity-planning method is based on theoretical calculations; an introduction to these calculations is given in the following subsections.

2.12.1 Grade of Service

How happy subscribers are depends on the *grade of service* (GoS; availability or quality of the service). The GoS depends on the network capacity that should meet the service demand of the customers. System faults, error rates, and other quality measures are not considered here. The most important factor in our study is if the call is successful or if it is blocked (see Figure 2.24), so we concentrate only on the evaluation of the blocking probability. For the probability of unsuccessful calls operators define the target value, the highest

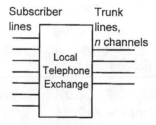

Figure 2.24 Local exchange and blocking. The purpose and principles of traffic engineering. Theoretical traffic engineering is needed to find the optimum (minimum) capacity of the network (e.g., exchanges, transmission lines, and radio channels) for a given grade of service. Traffic calculations are based on the busy-hour traffic intensity. For these calculations the traffic unit Erlangs is used. Examples: a subscriber line is used 6 min in an hour and traffic density is 100 mErl (maximum intensity is 1 Erl). Each channel of a 30-channel circuit is used on average 12 min in an hour and traffic intensity is 6 Erl (maximum intensity is 30 Erl).

probability of an unsuccessful call that they assume to be acceptable for their customers. The smaller this probability is, the more capacity they must build into the network.

Another factor we could use to define GoS is how long a length of time the subscriber has to wait until the service becomes available. We could design the network to keep customers in queue until, for example, a transmission channel becomes free. This waiting or queueing time is also essential to those who plan the telephone service where a person answers to incoming calls (e.g., switchboard service of an enterprise or customer service telephone).

2.12.2 Busy Hour

Network capacity planning is based on the so-called busy-hour traffic intensity and on other times when the GoS is better. "Busy hour" is an hour in the year when the average traffic intensity gets the highest value. To be accurate, the busy hour is determined by first selecting the 10 working days in a year with the highest traffic intensity; four consecutive 15-min periods (of those 10 days) with the highest traffic intensity make up the busy hour.

Figure 2.24 shows a summary of the information we need in the following analysis. The basic goal is to find a minimum capacity that gives the defined GoS. One example in the figure is a local exchange with a number of subscribers and a much smaller number N of trunk lines to the next exchange. If more than N subscribers have an external call at the same time, some of them are blocked, and must try again.

2.12.3 Traffic Density and the Erlang

The measure of traffic intensity is called *Erlang*, in honor of the Danish mathematician A. K. Erlang, the founder of the traffic theory; see Figure 2.23. The unit Erlang may be defined as:

- A unit of telephone traffic specifying the percentage of average use of a line or circuit (one channel);
- The ratio of time during which a circuit is occupied and the time for which the circuit is available to be occupied. A traffic that occupies a circuit for one hour during a busy hour is equal to 1 Erlang.

For example,

- If the traffic intensity of a subscriber line is 1 Erlang, it occupies 60 minutes in an hour.
- If a subscriber line is in use for 6 min in an hour (on average), the traffic intensity is 100 mErl.

- Maximum traffic intensity of a 2-Mbit/s (30 PCM channels) line system is 30 Erl, that is, all the channels are in use for 60 min during the busy hour.

The typical average busy-hour traffic volume generated by one subscriber is in the range of 10 mErl to 200 mErl. Low values are typical for residential use and high values for business subscribers.

2.12.4 Probability of Blocking

The problem in traffic engineering is determining the capacity if the average *offered traffic* intensity is known (or estimated). Offered traffic means the average generated total traffic including the traffic that is blocked in the system. Clearly the capacity should (at least usually) be higher than offered traffic, otherwise, many users would not be able to get service because all the lines would be occupied all the same time (on average), see the example of blocking in the local exchange in Figure 2.23. If all the trunk lines are occupied, new users are blocked, they receive a busy tone, and they have to try again. The essential question is: how much higher should the capacity be for the subscribers to feel that the GoS is acceptable?

The starting point is how often subscribers are allowed to be blocked and receive a busy tone. This probability of blockage for an acceptable GoS is usually set to be in the range of 0.2% to 5%, which means that every 500th to 20th call is blocked. When the average traffic load is estimated to increase into a certain volume, the network operator should increase the network capacity in order to keep the blocking probability below the defined GoS level.

The Poisson distribution is used as a probability model for these calculations. There are many ways to calculate the blockage probability, but the one we use here is known as "Molina lost calls held trunking formula," which is presented in Figure 2.25.

The formula that is given in Figure 2.25 is based on following assumptions [2]:

- Poisson arrival rate; Poisson-distributed call attempts;
- Equal traffic volume per source;
- Lost calls held; calls that are blocked stay in the system for an average holding time;
- Infinite number of sources; if some sources are blocked, this does not affect the total offered traffic and the total offered traffic by all sources together is taken as an input number.

Blocking occurs if a subscriber tries to make a call when all circuits (radio channels, trunk lines or maximum number of the switched circuits through an exchange) are occupied.

Probability model (Poisson distribution):

$$GoS = P(x{>}{=}n) = 1 - \sum_{x=0}^{n-1} \frac{A^x e^{-A}}{x!}$$

Where:
P = Probability of blockage
n = Number of servers (circuits)
A = Total offered traffic in Erlangs
e = 2.718281828
x! = Factorial of x (=1*2*...*x)

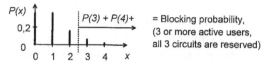

P(3) + P(4)+ = Blocking probability, (3 or more active users, all 3 circuits are reserved)

Figure 2.25 Probability of blocking. Example: A = 1 Erl (during the busy hour there is on average one active user); n = 3 (three channels are available for all the users); the probability that there is three or more users are active (all three available circuits are occupied) at a time and a new user is blocked is P(3) + P(4) + . . . = 1 − P(0) − P(1) − P(2) = 1 − 0,37 − 0,37 − 0,18 = 8%. This means that every twelfth call that a user makes is blocked and busy signal is received.

To find out the GoS, that is, blocking probability, we compute (using a Poisson distribution) the probability that not one channel is free when a subscriber makes a call. For this we take a value of the average (total) offered traffic as A and calculate the probability that traffic occupies all n channels or is even higher at that point in time (offered traffic load may be higher than n even though actual traffic never can exceed n). We get this by subtracting from one the probability that traffic is smaller than n.

Figure 2.24 gives one example of how the blocking probability is calculated. We actually calculate the probability that all n trunks are occupied or there is even more traffic. We get this by computing the probabilities, using the Poisson distribution, that all channels are free (x = 0), one channel is occupied (x = 1), or two channels are occupied (x = 2). We subtract the sum of these probabilities from one and get the blocking probability, that is, the probability that the number of occupied channels x ≥ 3. We get the result when there are three channels available and average offered traffic is 1 Erl (i.e., on average, one call on all the time) that the blocking probability is 8%. This means that approximately every 12th call is not successful.

Let us consider another example. The total offered average traffic during the busy hour is 2 Erl (A = 2); and the number of servers, for example transmission lines, is 5 (n = 5). Then the probability of blockage is 5.3% (P = 0.053). This means that, on average, during the busy hour, every 19th call is blocked, the busy tone is heard, and the subscriber has to redial.

When the number of channels or servers n is high, precalculated tables like Table 2.1 are used for network planning. It gives the required number of servers n when the GoS (= blocking probability in our study) and estimated offered traffic density A is given. For example, if the GoS is set to be 2% and offered traffic is 5 Erl (e.g., 100 subscribers with average traffic density of 50 mErl per subscriber) the network capacity should support at least 10 simultaneous calls (n = 10). If the capacity is smaller, for example n = 9, we get 2% with offered traffic 4.34 Erl, an offered traffic of 5 Erl would give a higher blocking rate.

When we look at Table 2.1 we note that when the number of servers is small, offered traffic intensity is in the order of one-tenth of the maximum traffic density. For example, with two channels, offered traffic intensity at blocking probability 1% is only 150 mErl. Both lines are allowed to be occupied on average only 9 min in an hour!

When the number of servers is high, allowed average traffic intensity is close to the maximum or even higher. Even when most of the channels are occupied, there are still some free for new calls because the number of channels is large.

Note that when high blocking probability is allowed, offered traffic may be higher than the number of available channels. A part of offered traffic is blocked and actual traffic, that part which is not blocked, naturally never gets a higher value than the number of channels in Erlangs.

Table 2.1 is calculated with a little bit more complicated formula that assumes that blocked calls are immediately cleared. This Erlang B formula gives slightly more optimistic results than the Poisson formula that we presented previously. This formula is used in Europe and Poisson is used in the United States for network planning. In order to compare the results of these two different approaches we consider an example where average offered traffic A = 2 Erl. If the number of circuits n = 5, we get blocking probability P = 0.0367 according to the Erlang B formula instead of 0.053, which we got previously with the Poisson or "Molina lost calls held" formula. Erlang B formula is [3]

$$P = \frac{A^n / n!}{\sum_{x=0}^{x=n} A^x / x!} \qquad (2.A)$$

Table 2.1
A Table for Network-Capacity Planning

n	Grade of service							
	0.5% E	1.0% E	2.0% E	3.0% E	5.0% E	10% E	20% E	50% E
1	0.01	0.01	0.02	0.03	0.05	0.11	0.25	1.00
2	0.11	0.15	0.22	0.28	0.38	0.60	1.00	2.73
3	0.35	0.46	0.60	0.72	0.90	1.27	1.93	4.59
4	0.70	0.87	1.09	1.26	1.52	2.05	2.95	6.50
5	1.13	1.36	1.66	1.88	2.22	2.88	4.01	8.44
6	1.62	1.91	2.28	2.54	2.96	3.76	5.11	10.4
7	2.16	2.50	2.94	3.25	3.74	4.67	6.23	12.4
8	2.73	3.13	3.63	3.99	4.54	5.60	7.37	14.3
9	3.33	3.78	4.34	4.75	5.37	6.55	8.53	16.3
10	3.96	4.46	5.08	5.53	6.22	7.51	9.69	18.3
12	5.28	5.88	6.61	7.14	7.95	9.47	12.0	22.2
15	7.38	8.11	9.01	9.65	10.6	12.5	15.6	28.2
20	11.1	12.0	13.2	14.0	15.3	17.6	21.6	38.2
25	15.0	16.1	17.5	18.5	20.0	22.8	27.7	48.1
30	19.0	20.3	21.9	23.1	24.8	28.1	33.8	58.1
35	23.2	24.6	26.4	27.7	29.7	33.4	40.0	68.1
40	27.4	29.0	31.0	32.4	34.6	38.8	46.2	78.1
45	31.7	33.4	35.6	37.2	39.6	44.2	52.3	88.1
50	36.0	37.9	40.3	41.9	44.5	49.6	58.5	98.1
55	40.4	42.4	44.9	46.7	49.5	55.0	64.7	108.1
60	44.8	46.9	49.6	51.6	54.6	60.4	70.9	118.1
65	49.2	51.5	54.4	56.4	59.6	65.8	77.1	128.1
70	53.7	56.1	59.1	61.3	64.7	71.3	83.3	138.1
75	58.2	60.7	63.9	66.2	69.7	76.7	89.5	148.1
80	62.7	65.4	68.7	71.1	74.8	82.2	95.8	158.1
85	67.2	70.0	73.5	76.0	79.9	87.7	102.0	168.0
90	71.8	74.7	78.3	80.9	85.0	93.2	108.2	178.0
95	76.3	79.4	83.1	85.9	90.1	98.6	114.4	188.0
100	80.9	84.1	88.0	90.8	95.2	104.1	120.6	198.0
110	90.1	93.5	97.7	100.7	105.5	115.1	133.1	218.0
140	118.0	122.0	127.0	130.6	136.4	148.1	170.5	278.0
200	174.6	179.7	186.2					
300	270.4	277.1	285.7					
400	367.2	375.2	385.9					
500	464.5	474.0	486.4					

where A is the offered traffic and n is the number of servers, for example, channels.

2.13 Problems and Review Questions

Problem 2.1: Describe how dialed digits are transferred from a subscriber's telephone to the local exchange.

Problem 2.2: Explain how the telephone attenuates the speaker's voice from the microphone to the earphone. *Hint*: Draw the current from the microphone in Figure 2.7 and imagine what happens to the magnetic field in the iron core of the transformer.

Problem 2.3: What is a 2W/4W hybrid and why is it needed at the end of the subscriber line?

Problem 2.4: Explain how a 2W/4W hybrid prevents the signal from the network (receive pair) from looping back to the transmit pair.

Problem 2.5: Explain the basic principle of the telephone call routing through the switching hierarchy to another region of the country.

Problem 2.6: A network has N subscribers. Each subscriber is connected directly to all other subscribers.

    a. What is the total number of bidirectional lines L of the network?

    b. What is the value of L for N = 2, 10, 100, and 1000?

    c. How many lines must be built to each subscriber?

    d. Is this kind of network structure suitable for a public telecommunications network? Explain.

Problem 2.7: What are the basic differences between the public and private telecommunications networks? List a few examples of both public and private networks.

Problem 2.8: What is ISDN? How does the service and structure of a subscriber interface differ from the conventional analog telephone service?

Problem 2.9: How does IN differ from an ordinary fixed telephone network? List some examples of IN services.

Problem 2.10: A PBX/PABX has seven user channels to a public exchange. During the busy hour, on average, 3.4 lines are occupied.

    a. What is the traffic intensity during the busy hour?

    b. Estimate, with the help of the Table 2.1, the GoS (blocking probability).

Problem 2.11: What is the total offered traffic intensity from a PBX/PABX to a PSTN if there are ten calls, with a duration of 6 min each, during one hour?

Problem 2.12: A subscriber makes one 6-min call in one day between 10.00 and 10.06 o'clock. What is the average traffic intensity of his/her subscriber line during (a) 10.00 to 10.06, (b) 10.00 to 10.15, (c) 10.00 to 11.00, and (d) 00.00 to 24.00 of that day?

Problem 2.13: Calculate the blocking probability (GoS) when the total offered traffic is 2 Erl and the number of available transmission channels in the network is 5.

Problem 2.14: Draw two curves for GoS levels 1% and 10%. Use the vertical axis as a ratio A/n from 1% to 100% and the horizontal axis as a number of circuits N from 1 to 20. Use traffic engineering Table 2.1. What can you say about network utilization when the number of circuits n is small? How does the utilization of the circuits depend on the required probability of blocking P?

Problem 2.15: What should be the approximate capacity of the network (how many channels should be available) if there are 100 subscribers and each of them generates offered traffic of 40 mErl. Probability of blocking is (a) 20% and (b) 1%. Use the traffic engineering Table 2.1.

Problem 2.16: There are 20 users of a keyphone system that has two lines to a public network. What is the blocking probability when each user generates 100-mErl offered traffic?

Problem 2.17: A keyphone system with three lines to the local exchange is used in an office of 10 persons. Each of them uses the phone for an external call of 15 min in a busy hour. How many lines are reserved on average during an hour? What is the blocking probability? What do you think about the capacity of this system?

Problem 2.18: Subscribers of a local exchange generate 100 mErl of traffic through the exchange to the network. What should the number of trunk channels be if the number of subscribers in the area is (a) 10, (b) 100, (c) 1000, and (d) 4000? The allowed blocking level is 1%. Use Table 2.1 to estimate the required number of circuits.

References

[1]   Bellcore, *Telecommunications Transmission Engineering*, Bellcore Technical Publications, 1990.

[2]   Freeman, L. F., *Telecommunication System Engineering*, New York, NY: John Wiley & Sons, 1996.

3

Signals Carried Over the Network

Services that the telecommunications networks provide have different characteristics. Required characteristics depend on the applications we use. In order to meet these different requirements, many different network technologies that are optimized for each type of service are in use. Even new technologies are needed and being put into use, a current example of which is *asynchronous transfer mode* (ATM). In order to understand the present structure of the telecommunications network, we need to get a view of what kinds of signals are transmitted through the telecommunications network and what are their requirements. In this chapter we look at the requirements of various applications, characteristics of analog voice channels, fundamental differences of analog and digital signals, analog-to-digital conversion, and a measure of decibel signal level.

3.1 Types of Information and Their Requirements

Modern digital networks transmit digital information transparently, that is, the network does not necessarily need to know information on the content of the data. The information that is transmitted through the network may be:

- Speech (telephony);
- Moving images (television or video);
- Printed pages or still picture (fax or facsimile);
- Text (electronic mail);
- Music;
- All types of computer information such as program files.

However, although all information is coded in digital form, the transmission requirements are highly dependent on the application; and because of these different requirements, we have different networks and technologies in use. Video and email applications, for example, require different architectures. There have been two main paths in the development of network technologies: one for speech services and another for data services. The telephone network and ISDN have been developed for constant bit-rate voice communication that is well-suited to speech transmission. Data networks such as X.25 and LANs have been developed for bursty data transmission.

The constant bit-rate requirement for speech follows from the principle that voice signals are transmitted in digital form as samples at regular intervals, as we will see in Section 3.6. Data transmission is bursty by nature. Sometimes we may copy a file across the network, but most of the time we work locally with our workstation.

When many different applications are integrated (multimedia) into data communication environments, both basic types of service requirements of constant bit-rate voice and bursty data must be fulfilled and we need a concept that is able to meet both types of requirements. ATM is a network technology that was developed to be suitable and efficient for transferring all types of information.

In Figure 3.1 different applications are compared from the communication requirements point of view. The applications are ordinary speech; *computer-aided design* (CAD), (a service where high-resolution graphical information is transmitted); moving images (video); file transfer; and multimedia with integrated video, voice, and data. The importance of the transmission requirements for each application is explained in the following:

Transmission Characteristics	Voice Service	CAD Service	Video Service	File Transfer Service	Interactive Multimedia
Bandwidth requirement	Low, fixed	Very high, variable	Very high, fixed	High, variable	High, variable
Data loss tolerance	Tolerant	Non-tolerant	Tolerant	Non-tolerant	Tolerant or non-tolerant
Fixed delay tolerance	Low delay	Tolerant	Tolerant	Tolerant	Low delay
Variable delay tolerance	No	Tolerant	No	Tolerant	No
Peak information rate	Fixed	Very high	Fixed	High	Very high

CAD = Computer Aided Design

Figure 3.1 Communication requirements of different applications.

Data Rate or Bandwidth Requirement

Graphical work station communication requires temporary high bit-rate transmission, and video transmission requires continuous high data-rate transmission over the network. High-resolution graphical figures and moving images usually represent large amounts of data, which are transmitted over the network in a CAD application only when an update of the graphics is needed, but in the case of video, transmission is continuous. Instead of *data rate* we sometimes use the term *bandwidth* because they are closely related to each other, as we will see in Chapter 4.

Data Loss Tolerance

Noise and other disturbances in the network may cause errors in the transmitted data. If errors occur, some amount of data may be lost. Conventional voice and video transmission services are used by human beings, and they can tolerate accidental short disturbances. In computer communications a single erroneous bit usually destroys a whole data frame, which may contain a large amount of data. The loss of one frame destroys the transmission of a large file that is transferred in multiple frames. Most of the data transmission systems are able to retransmit faulty data frames. However, when errors occur, there is still a small probability that recovery procedures may fail and some data will be lost.

Fixed Delay Tolerance

When communication is interactive, as voice communication usually is, the two-way transmission delay should be very short for good quality. In the case of voice, it should be in the order of some tens of milliseconds. Otherwise we feel that quality is degraded because the response from the other party is delayed. We tolerate much longer delays in the case of ordinary data applications when we are awaiting the response to our "click" command.

Variable Delay Tolerance

Voice and moving video information is transmitted as samples at regular periods of time. The reconstruction of images and voice requires that all sample values be received sequentially and suffer the same delay. Conventional data networks recover from errors with the help of retransmission of the frames in error. This is a very efficient error recovery scheme, but it introduces some additional and variable delay. For voice applications this variable delay is often a worse choice than some lost data.

Peak Information Rate

The information rate is usually constant for voice and video. Values of the samples with constant length contain voice or video information and are transmitted at a constant rate. In data communication applications we usually work locally, and every now and then a high data rate is needed to load graphical information or files. A peak load is typically in the order of a thousand times higher than the average transmission capacity we use.

3.2 Simple, Half-Duplex, and Full-Duplex Communication

In telecommunication systems the transmission of information may be unidirectional or bidirectional. The unidirectional systems that transmit in one direction only are called *simplex,* and the bidirectional systems that are able to transmit in both directions are called *duplex* systems. We can implement bidirectional information transfer with *half-* or *full-duplex* transmission as shown in Figure 3.2.

In simplex operation the signal is transmitted in one direction only. An example of this principle is broadcast television, where TV signals are sent from a transmitter to TV sets only and not in the other direction. Another example is a paging system that allows a user only to receive alphanumerical messages.

In half-duplex operation the signal is transmitted in both directions but only in one direction at a time. An example of this is a mobile radio system where the person speaking must indicate by the word "over" when the other person is allowed to transmit. LANs use a high-speed half-duplex transmission over the cable even though users may feel that the communication is continuously bidirectional, that is, full duplex.

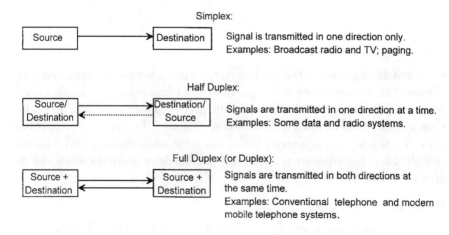

Figure 3.2 Simplex, half-duplex, and full-duplex transmission.

In full duplex, or simply duplex, operation signals are transmitted in both directions at the same time. An example of this is an ordinary telephone conversation where it is possible for both people to speak simultaneously. Most modern telecommunication systems use the full duplex principle, which we call duplex operation, for short.

3.3 Frequency and Bandwidth

In order to understand the requirements of different applications for a telecommunications network, it is fundamental to understand the concepts of frequency and bandwidth. The information that we transmit through a telecommunications network, whether it is analog or digital, is in the form of electrical voltage or current. The value of this voltage or current changes through time, and this alteration contains information.

The transmitted signal (the alteration of voltage or current) consists of multiple frequencies. The range of frequencies is called the bandwidth of the signal. The bandwidth is one of the most important characteristics of analog information and the most important limiting factor for the data rate of digital information transfer.

3.3.1 Frequency

We can see the telecommunication signal as a combination of many cosine waves with different strengths and frequencies. The frequency means how many cycles the wave oscillates in a second. As an example of the concept of frequency, we hear the oscillation of air pressure as sound. We are able to hear frequencies approximately in the range of 20 Hz to 15 kHz, where Hz (Hertz) represents the number of cycles in a second. An example of the different frequencies is the keys of a piano. The right hand side keys generate basic frequencies in the order of a thousand Hertz per second and the left-hand side keys in the order of a hundred Hertz per second.

In electrical terms, an *alternating current* (AC) changes its direction of flow several times a second. This variation in direction is known as a "cycle," and the term "frequency" refers to the number of cycles in a second that is measured in Hertz. If a signal has 1000 complete cycles in a second, then its frequency is 1000 Hz or 1 kHz. A pure sine wave in Figure 3.3 is generated with a loop of wire rotated in a magnetic field at a constant rate. This fundamental waveform can be seen as a cosine of the angle of the phasor rotating at a constant rate. The strength of the voltage or current alters according to the cosine curve when time increases. The length of the phasor corresponds to the

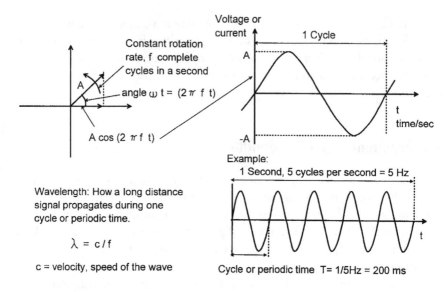

Figure 3.3 Cosine wave and frequency. The example shows 1 sec, 5 cycles/s = 5 Hz and cycle or periodic time $T = 1/5$ Hz = 200 ms. The wavelength represents how a long-distance signal propagates during one cycle or periodic time and is denoted $\lambda = c/f$, where c is the velocity or speed of the wave.

maximum value of the signal and is called *amplitude*, shown as A in Figure 3.3.

We can consider any telecommunication signal as a sum of these funda-mental waveform cosine waves that are expressed as

$$v(t) = A \cos(\omega t) = A \cos(2\pi f t) \qquad (3.A)$$

where f is frequency, the number of complete cycles in a second expressed in Hertz, 1 Hz = $1/s$; t is time in seconds; and ω is angular frequency in radians per second. Then the angular frequency is $\omega = 2\pi f$, which means that one complete cycle of a phasor makes up an angle of 2π radians. Phase, i.e., angle at time instant $t = 0$ is assumed to be 0 in the formula 3.A.

The *periodic time* or *period* T in seconds represents the time of one complete cycle

$$T = 1/f \quad \text{and} \quad f = 1/T \qquad (3.B)$$

The wavelength λ represents the propagation distance in one cycle time and, thus,

$$\lambda = c/f = cT \qquad\qquad (3.C)$$

where c is the velocity of the signal. For sound waves the velocity in the air is approximately $c = 346$ m/s; and for light or radio waves, approximately, $c = 300{,}000$ km/s.

3.3.2 Bandwidth

The voice signal, which is the most common message in telecommunications network, does not look similar to a pure cosine wave in Figure 3.3. It contains many cosine waves with different frequencies and amplitudes combined together. The range of frequencies that is needed for good enough quality voice, so that the speaker can be recognized, was defined to be the range from 300 Hz to 3400 Hz. This means that the bandwidth of the telephone channel through the network is 3400 Hz − 300 Hz = 3.1 kHz, as shown in Figure 3.4. A human voice contains much higher frequencies, but this bandwidth was defined as a compromise between quality and cost. It is wide enough to recognize the speaker.

Bandwidth is not strictly limited in practice, but signal attenuation increases heavily at the lower and upper cut-off frequencies, 300 Hz and 3.4 Hz, of the voice band. The bandwidth is normally measured from the

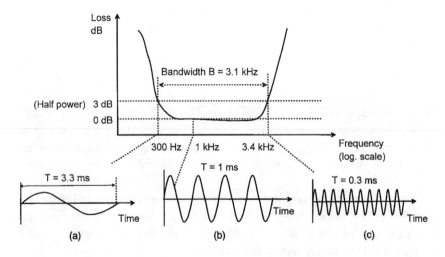

Figure 3.4 Bandwidth of the telephone speech channel. The attenuation or loss of 3 dB decreases power to half and causes a corresponding voltage drop from 1 to 0.707. Sine wave oscillating (a) 300 times/s attenuates to half power, (b) 1000 times/s does not attenuate at all, and (c) 3400 times/s atenuates to half power.

points where the signal power is dropped to half from its maximum power. Attenuation or loss of channel is given as a logarithmic measure called the *decibel* (dB), and half-power points correspond to a 3-dB loss. Decibels are discussed later in this chapter.

Bandwidth, together with noise, is the major factor that determines the information-carrying capacity of a telecommunication channel. The term "bandwidth" is often used instead of data rate because they are closely related, as we will see in Chapter 4.

3.4 Analog and Digital Signals and Systems

Most of the systems in the modern telecommunications network are digital rather than analog. In this section we look at the fundamental characteristics of analog and digital signals and how they influence the performance and operation of telecommunication systems.

3.4.1 Analog and Digital Signals

The difference between analog and digital forms is easily understood by looking at the two watches in Figure 3.5. A true analog watch has hands that are constantly moving and always show the exact time. A digital watch displays "digits," and the display jumps from second to second and shows only discrete values of time.

Another example could be the slope of analog voltage where all values of voltage can be measured as shown in Figure 3.5. In "digital slope" only discrete values may be measured. In the example of the figure, we have eight discrete values, 0 to 7, in the digital slope. This does not mean that the digital systems have worse performance than analog systems. If we want to improve the accuracy of the digital system, we just increase the number of steps and, in principle, any voltage level can be represented with the digital system as well.

A special and very important case of *digital signals* is a binary signal where only two values, binary digits "0" and "1", are present, as illustrated in Figure 3.5. Examples of binary signals include light on and off, voltage versus no voltage, and low current versus high current.

Binary signals are used internally in computers and other digital systems to represent any digital signal. For example, we can encode eight voltage levels of the slope in Figure 3.5 into three binary bits that then represent one of the $2^3 = 8$ (0 to 7) different values. As another example, a digital signal with 8-bit words or bytes (often called *octets* in digital telecommunication systems) can

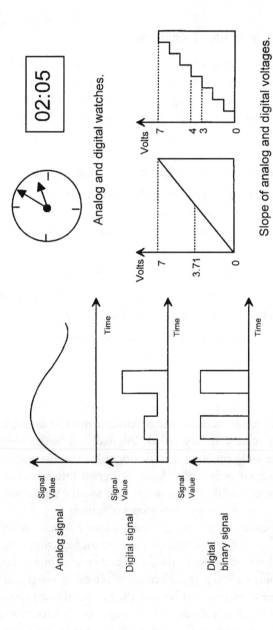

Analog and digital watches.

Slope of analog and digital voltages.

Figure 3.5 Analog and digital signals.

represent 2^8 = 256 discrete values of a digital signal. This kind of digital signal is used to represent analog voice, where each sample of a voice signal is encoded into 8-bit words, as we will explain in Section 3.6.

3.4.2   Advantages of Digital Technology

Analog systems in a telecommunications network have gradually been replaced with digital systems. Digital systems have several advantages that make them much more attractive than analog systems now that digital technology is available. The most important advantages of digital technology over analog technology are:

- Digital functions make a high scale of integration possible.
- Digital technology means lower cost, better reliability, less floor space, and less power consumption.
- Digital technology means that communication quality is independent of distance.
- Digital technology means better noise tolerance.
- Digital network is ideal for growing data communications.
- Digital technology makes new services available.
- A digital system has a high transmission capacity.
- A digital network offers flexibility.

An analog system requires the accurate detection of signal values inside its dynamic range, that is, between the maximum and minimum values of the signal. Digital systems use binary signals internally. A binary signal has only two values, and the only problem is to distinguish these two values from each other. The dynamic range is well defined and good linearity is not required. This makes the elements of digital circuits simple and the utilization of compact technology, digital circuit integration becomes feasible.

As a consequence, circuit integration leads to a smaller number of electronic components, smaller equipment, lower manufacturing cost, and lower maintenance cost because of better reliability and smaller power consumption. More and more complex integrated circuits replace many lower scale integrated circuits. This decreases system cost because the increased complexity of components does not cost much in volume. When integrated circuits are manufactured in volumes, complex circuits do not cost much more than less complex circuits. In addition, the smaller number of separate components gives better reliability.

In long-distance connections, we have to amplify or regenerate the signal on the line many times. When we amplify an analog signal on the line, we amplify noise at the same time. This added noise decreases the quality of an analog signal, that is, decreases the *signal-to-noise* (S/N) ratio.

In the case of a digital system we use regenerators or repeaters instead of amplifiers. A repeater regenerates the signal symbol by symbol, that is, transmits further the value that is closest to the received value. The regenerated signal is a sequence of digital symbols with nominal values and, thus, contains no noise. If the noise is low enough in the input of each regenerator, symbols of the digital signal are regenerated without errors and we receive exactly the same digital message on the other side of the world as it was at the transmitting end. The operation of a digital repeater or regenerator is described in Chapter 4.

Modern switches digitize speech in the subscriber interface. If the path through the network is fully digital, conversion back to analog is done only at the far end. There is only one analog-to-digital and one digital-to-analog conversion regardless of the communication distance, that is, whether we make a call to our neighbor or to other side of the world.

Digital systems have to identify only a set of discrete values. If symbols are not mixed because of a too high noise level, noise does not have any impact on the operation. Analog communication usually requires a much better S/N ratio than digital communication. As a consequence, digital systems can utilize channels with a much higher noise level than analog systems.

If the network is analog, a digital message has to be modulated into the frequency band of the analog telecommunication channel. This reduces the capacity available for the user. For example, a voice channel in the digital telephone network has a data capacity of 64 kbit/s. If we use it via an analog interface analog subscriber loop, the data rate is restricted in practice to approximately 30 kbit/s. With a digital subscriber line, which is used in ISDN, the user data rate is higher and exactly the same 64 kbit/s that is used inside the network.

Digital systems are ideal for software control because digital circuits operate in a numerical way. Integrated software makes systems flexible, and new functions needed for new services are easier to update. IN services that we reviewed in Section 2.10 are good examples of these new services. As another example, we would not have cellular telephone service without digital software-controlled systems in the network.

The digital processing of information makes better utilization of channels possible; for example, several digital broadcast radio channels fit into a band of one analog broadcast radio channel. In Chapter 4 we will see that digital signals tolerate higher disturbances than analog signals—this is one reason behind the better frequency efficiency. Low-cost multiplexing (no analog filtering and modulation circuitry required) and efficient use of optical transmission media make high-capacity digital systems feasible. Optical systems today transmit digital signals as a series of short light pulses. The distortion of these digital pulses does not influence the quality of the message because distorted pulses are regenerated, which eliminates distortion.

All types of analog signals can be converted into digital signals. When this is done, the digital network is able to carry any information. Bits are handled in the same way whether they represent voice or data.

Analog systems are different for each application because of different performance requirements. For example, a telephone connection requires channels with approximately 4-kHz bandwidth, but television signals require 5-MHz bandwidth with much better S/N. In digital systems the corresponding characteristic is the data rate. For example, an analog telephone signal requires 64 kbit/s and video with much wider bandwidth requires 2 to 140 Mbit/s depending on the coding scheme in use. We can use one high data-rate system for a single video channel or a large number of speech channels.

The digital technology provides efficient multiplexing for sharing capacity in high data-rate connections. This makes high capacity digital networks and systems flexible. The same system, if it provides a high enough data rate, can be used for any application.

3.4.3 Examples of Messages

In the previous sections we described the characteristics of the digital and analog signals and systems. Now we look at some simple examples of information sources that produce messages that are transmitted through the network. There are many different information sources, including machines as well as people, and messages or signals appear in various forms. As for signals we can identify the two main distinct message categories: analog and digital.

3.4.3.1 Information, Messages, and Signals

A concept of information is central to communication. However, information is a loaded word, implying schematic and philosophical notions; and therefore, we prefer to use the word message instead. A message means the physical manifestation of information produced by a source. Systems handling messages convert them into electrical signals suitable, for example, for a certain transmission media.

3.4.3.2 Analog Message

An analog message is a physical quantity that varies through time, usually in a smooth and continuous fashion. Examples of analog messages are acoustic pressure produced when you speak or light intensity at one point in an analog television image. One example of an analog message is the voice current on a conventional subscriber telephone line, as illustrated in Figure 3.6. In Section 2.2 we explained how the current is produced.

Since the information resides in a time-varying wave form, an analog communication system should deliver this waveform with a specific degree of fidelity. Because the strength of signals may vary between 30 dB to 100 dB, depending on the application, the analog systems should have good linearity from the weakest signal to 1000 to 10,000 million times stronger signal values.

3.4.3.3 Digital Message

A digital message is an ordered sequence of symbols selected from a finite set of discrete elements. Examples of digital messages are the letters printed on this page or the keys you press at a computer terminal. When we press a key at our computer terminal, each key stroke represents a digital message that is then encoded into a set of bits for binary transmission.

Since the information resides in discrete symbols, a digital communication system should deliver these symbols with a specified degree of accuracy in a specified amount of time. The main concern in the system design is that symbols remain unchanged, which is the final requirement for transmission accuracy.

We need modems to transmit digital messages over analog channels. The modems receive a message from the terminal in the form of binary data and send it as an analog waveform to the speech channel, as shown in Figure 3.6. High data-rate modems do not modulate, that is, change the analog waveform, at the rate of the binary data they receive from the terminal. Instead they encode a set of bits into a digital symbol that may get many values. These multilevel symbols contain a set of bits and change the analog signal on the line. This increases the data rate through the speech channel, which has quite a narrow bandwidth, as we will see in Chapter 4.

When a digital network is used to transmit digital messages, signals are usually binary from end to end. Instead of a modem, a network terminal is needed at the subscriber's premises to encode digital pulses into a suitable form for cable transmission to an exchange site; see example of ISDN in Figure 3.6.

3.5 Analog Signals Over Digital Networks

In this section we look at how analog signals are handled before transmission through a digital network. In the next section we concentrate on the PCM

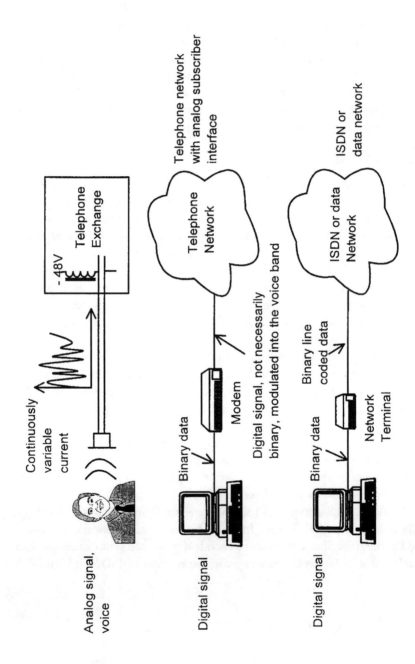

Figure 3.6 Examples of messages.

that is performed in the network on our voice during a telephone call, and in Section 3.7 we present a brief review of other voice-coding schemes.

If a digital signal is to be transmitted through an analog network, it has to be converted into an analog signal suitable for the frequency band of the channel, as we saw in Figure 3.6. Digital networks provide communication only with a set of discrete symbols (in the binary case these symbols are called bits) at a certain data rate and the analog signal has to be converted into a series of these symbols for digital communication. The data rate of a digital network corresponds to the channel bandwidth of an analog network. The higher the data rate, the wider the bandwidth that is usually required and vice versa.

If the network is fully digital, analog voice is encoded into a digital form at the transmitting end and decoded into analog form at the receiving end, as shown in Figure 3.7. This coding is performed in the subscriber interface of a digital telephone exchange and, in the case of ISDN service, in the subscriber's ISDN telephone.

The main phases of this process, as shown in Figure 3.7, are:

- *Analog-to-digital conversion* (A/D): An analog signal is sampled at the sampling frequency and the sample values are then represented as numerical values by the encoder. These values, presented as binary words, are then transmitted within regular time periods through the digital channel.

- *Digital-to-analog conversion* (D/A): At the other end of the channel the decoder receives numerical values of the samples that indicate the

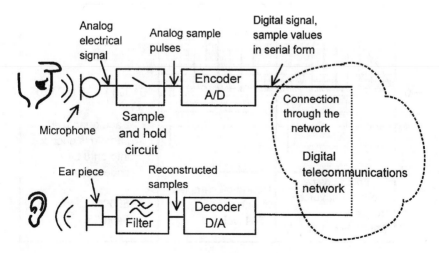

Figure 3.7 Analog voice signal through a digital network.

values of the analog signal at certain points of time. The sample pulses that have amplitudes corresponding to the values of the original signal at sampling instants are reconstructed and the series they form is filtered to produce an analog signal close to the original one.

The methods for these A/D and D/A conversions have to be specified in detail so that the reproduction of the analog signal is compatible with the production of the digital signal that may have occurred on the other side of the world. In the next section we describe the method that is used in the telecommunications network and internationally standardized by the *International Telecommunications Union* (ITU).

3.6 Pulse Code Modulation

PCM is a standardized method that is used in the telephone network to change an analog signal to a digital one for transmission through the digital telecommunications network. The analog signal is first sampled at a 8-kHz sampling rate; then each sample is quantized into one of 256 levels and then encoded into digital 8-bit words. This encoding process is illustrated in Figure 3.8. The overall data rate of one speech signal becomes 8000 × 8 = 64 kbit/s. This same data rate is available for data transmission through each

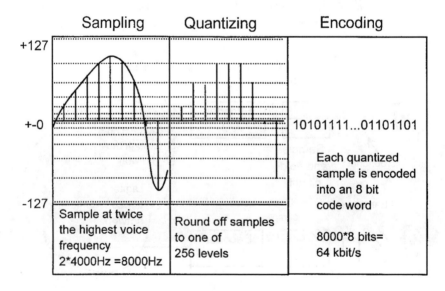

Figure 3.8 Pulse code modulation.

speech channel in the network. In the United States one bit of eight in every sixth frame is "robbed" for in-band signaling and the available transparent data capacity of a single speech channel in the network is reduced to 8000 × 7 = 56 kbit/s.

Now we take a more detailed look at the three main processing phases of the PCM in the telecommunications network. Note that this principle is employed by all systems when there is a need to process analog signals by a digital system. Sampling rates and the number of quantizing levels vary from application to application, but the basic principle and phases of the process remain the same.

3.6.1 Sampling

The amplitude of an analog signal is sampled first. The more samples per second there are, the more representative of the analog signal the samples will be. After sampling the signal value is known only at discrete points in time, called sampling instants. If these points have a sufficiently close spacing, a smooth curve drawn through them allows us to interpolate intermediate values to any degree of accuracy. We can therefore say that a continuous curve is adequately described by the sample values alone.

In a similar fashion, an electrical signal can be reproduced from an appropriate set of instantaneous samples. The number of samples per second is called the sampling *frequency* or sampling rate and it depends on the highest frequency component present in the analog signal. The relation of sampling frequency and the highest frequency of the signal to be sampled is stated as follows.

If the sampling frequency, f_s, is higher than two times the highest frequency component of the analog signal, W, the original analog signal is completely described by these instantaneous samples alone, that is, $f_s > 2W$.

This minimum sampling frequency is sometimes called the *Nyquist rate*. We can describe it in other words as: an analog signal with the highest frequency component as W Hz that is completely described by instantaneous sample values uniformly spaced in time within a period.

$$T_S = 1/f_S < 1/(2W) \tag{3.D}$$

Figure 3.9 represents the operation principle of a sampling circuit and an analog signal before and after sampling both in the time and frequency domains. The sampling circuit contains a generator, G, that produces short sampling pulses at the sampling frequency f_s. These sampling pulses close the switch of a relay at each sampling instant for a short period of time. The

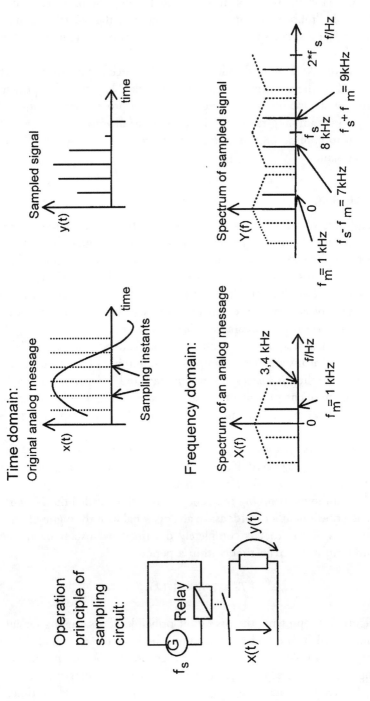

Figure 3.9 Sampling.

original analog signal $x(t)$ is sampled each time the switch is closed and a sampled signal $y(t)$ is produced. The sampled analog signal $y(t)$ contains short pulses that represent signal $x(t)$ values at discrete points in time. This sampling process that produces $y(t)$ is known as *pulse amplitude modulation* (PAM) because the amplitudes of the pulses contain the values of $x(t)$.

The time domain curves in the Figure 3.9 show the original continuous analog signal $x(t)$ and the sampled signal $y(t)$. The sampled signal $y(t)$ contains values of an analog signal at sampling instants. We can imagine that if the sampling frequency f_s is high, that is, the distance between sampling instants T_s is short, the sample pulses describe the original signal quite well. We could draw a line that connects the peak values of the pulses and the shape of this curve would be close to the original signal shape of $x(t)$.

The changes of $x(t)$ are related to the frequency content of $x(t)$. The more rapidly $x(t)$ changes, the higher frequency components it contains. This explains why the sampling frequency is related to the highest frequency that the analog signal contains. From the time domain figure we understand that the sampling frequency must be much higher than the highest frequency of the analog message. Otherwise, rapid changes of signal $x(t)$ between sampling instants could not be described by sample values. The accurate answer to how much higher it should be can be understood more easily via the frequency domain.

The frequency domain descriptions in Figure 3.9 show the spectrum of $x(t)$ and the sampled signal $y(t)$. Before sampling the spectrum $X(f)$ of $x(t)$ contains speech frequencies up to 3.4 kHz, shown as a dashed line in the figure. As an example of the frequency components of speech we drew the spectrum of 1-kHz cosine wave as a solid spectral line at the 1-kHz point on the frequency axis.

After sampling the spectrum of the message appears around the sampling frequency. If the message contains a single 1-kHz frequency component, after sampling we will have components at 1 kHz, 8 kHz − 1 kHz = 7 kHz and at 8 kHz + 1 kHz = 9 kHz, as seen in the figure. In addition to these components, sampling also generates components around double sampling frequency, three times sampling frequency, etc.

The reproduction of an original signal from a sampled signal is performed by a low-pass filter and in this case the bandwidth should be 4 kHz, that is, half the sampling frequency. We see that this filter would let through only a 1-kHz component of the spectrum in Figure 3.9. That is actually the original analog signal. With the help of the low pass filter we have successfully reproduced the original analog message from the samples alone.

If we increase the frequency of an analog message $x(t)$ from 1 kHz to 2 kHz we will have the lowest component of the sampled signal at 2 kHz, the solid spectral line at 1 kHz will move to the right, the next spectral component will be at 8 kHz − 2 kHz = 6 kHz, and the solid line at 7 kHz will move to the left. Low-pass filtering will still give the original 2-kHz message. Now if we increase the frequency beyond 4 kHz, say 5 kHz, we will get components at 5 kHz and 8 kHz − 5 kHz = 3 kHz and low-pass filtering will give a 3-kHz signal instead of the original 5-kHz signal. Reproduction will not work anymore because the frequency of an analog signal has exceeded half of the sampling frequency.

We have seen now that the sampling frequency must be more than twice the highest frequency component of the original signal to be encoded; otherwise, the message spectra around zero frequency and sampling frequency will overlap. This can be seen from the spectrum $Y(f)$ in Figure 3.9 if we imagine what happens if $W > f_s/2$. From the spectrum of the sampled signal $Y(f)$ in Figure 3.9, we also see that the message can be completely reconstructed from a PAM signal with a 4-kHz low-pass filter if $W < f_s/2$. This requirement is fundamental for all digital signal processing.

The highest frequency of voice that will be transmitted is chosen to be 3400 Hz, and the sampling frequency is standardized to be 8000 Hz, leaving enough guard band for filtering. Samples are then taken at intervals of $T_s = 125$ μs.

In the sampling process a PAM signal $y(t)$ is created. The amplitudes of PAM pulses follow the original analog signal. Note that the samples are still analog having any analog value between the minimum and maximum values of the original signal.

3.6.2 Quantizing

In the previous section we utilized sampling that produces a PAM signal representing discrete but still analog values of the original analog message at the sampling instants. In order to transmit the sample values via a digital system, we have to represent each sample value in a numerical form. This requires quantizing where each accurate sample value is rounded off to the closest numerical value in a set of digital words in use. Figure 3.10 represents the original and the quantized signals. The latter stays at the sample value until the next sampling instant.

In this quantizing process the information in accurate signal values is lost because of rounding off and the original signal cannot be reproduced exactly anymore. The quality of the coding depends on the number of quantum levels that is defined to provide the required performance. The more quantum levels we use, the better performance we get. For example, for voice signals 256 levels (8-bit binary words) are adequate, but for music encoding (CD-record

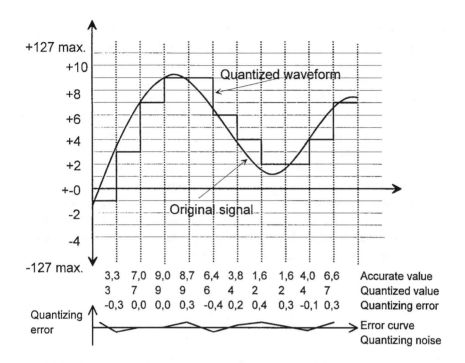

Figure 3.10 Quantizing.

player) 65536 levels (16-bit binary word) are needed to give sufficient performance.

In the case of binary coding, the number of quantum levels is $q = 2^n$, where q denotes the number of quantum levels and n, the length in bits of the binary code words that describe the sample values.

The better quality we require, the more quantum levels we need and the longer sample words we have to use. This leads to the requirement of a higher bit rate for transmission of the data representing the original message. The data rate must be so high that the digital word of the previous sample will be transmitted before the next one is available for transmission. In each system a certain compromise has to be made between quality and the data rate.

In uniform quantizing, the quantum levels are uniformly spaced between certain minimum and maximum values of the analog signal. In the next section we consider quantizing noise that the rounding off produces in the case of uniform quantizing.

3.6.3 Quantizing Noise

Quantizing causes signal distortion because the sample values no longer represent the accurate values of the analog signal. Usually this distortion caused by rounding off in quantizing is small compared to the signal. The maximum distortion, maximum quantizing error, is half of the distance between quantum levels. This distortion is heard and theoretically modeled as noise; see the quantizing error curve in Figure 3.10. We can imagine that the decoder first receives accurate sample values and produces a perfect original signal. Then the quantizing error is added on top of the perfect signal just as we hear, for example, background noise on top of an ideal voice or music signal.

The rounding off causes an error that is independent of the message because quantizing levels are close to each other and we can assume that the signal has the same probability to be anywhere between two levels at a certain sampling instant as shown in Figure 3.11. This error has a uniform probability density function and a zero mean. When we define the signal to have values between −1 and +1, it can be shown that the quantum noise power is equal to the variance of quantizing error and is given by

$$N = \sigma_q^2 = \frac{1}{3q^2} \qquad (3.E)$$

where $N = \sigma_q^2$ = quantization noise power and q = the number of quantum levels.

We see that if the number of quantum levels is increased, quantizing noise power decreases rapidly. We get the maximum *signal-to-quantizing noise ratio* (SQR) of linear quantizing when the maximum signal power is equal to one (the power is a square of the signal value that was defined to be between −1 and +1)

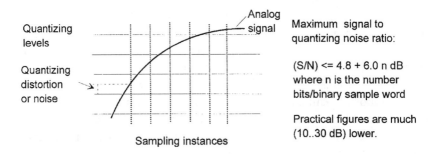

Maximum signal to quantizing noise ratio:

(S/N) <= 4.8 + 6.0 n dB where n is the number bits/binary sample word

Practical figures are much (10..30 dB) lower.

Figure 3.11 Quantizing noise and S/N. In the quantizing process accurate sample values are lost and we cannot reproduce a perfect original signal. The more levels (the more bits/sample, the higher bit rate) we use, the better performance we get, that is, higher S/N.

$$\left(\frac{S}{N}\right)_{\text{D}} = 3q^2 \tag{3.F}$$

where S = signal power, $N = \sigma_q^2$ = power of quantization noise, and q = number of quantum levels.

We can easily show further that in the case of linear quantizing and binary words, the absolute maximum S/N in decibels in the case of linear quantizing is

$$S/N \le 10 \log_{10}(3q^2) = 10 \log_{10}(3 \cdot 2^{2n}) = 4.8 + 6.0n \text{ dB} \tag{3.G}$$

where n = the number of bits per binary sample.

The logarithmic measure decibel is described at the end of this chapter. The preceding formula gives the absolute maximum S/N of a system that uses uniform quantizing and codes sample values into n-bit binary words.

However, we assumed that the average power of the analog signal equals the maximum power, that is, all the sample words have the maximum value. In practice, this cannot be the case and the average S/N is some tens of decibels lower than the maximum value given by the formula. How much lower an S/N we have in a practical system depends on the dynamic range that we reserve for the highest signal levels (the distance between the average signal power and the maximum signal power) in order to avoid the clipping of the signal and consequent severe distortion.

3.6.4 Nonuniform Quantizing

The goal in the coder design is to get as good an average S/N as possible when the sampling rate and the number of bits for each sample are given. Linear quantizing is not the optimum solution because at low signal levels the quantizing noise is high and the S/N is very low. At higher signal levels the quantizing noise is the same even though we would tolerate a high noise level. We should define quantizing levels in such a way that performance is acceptable over a wide dynamic range of the voice. This requires that quantum levels are not uniformly spaced and we call this nonuniform quantizing.

In nonuniform quantizing we use more code words and have a shorter distance between quantum levels for low-level samples and allow higher quantizing distortion at high-level samples. This is reasonable because higher noise is not so disturbing when the signal level is higher as well. To do this, we compress the voice signal in an encoder and expand it in a decoder. This expanding/compressing process is known as *companding* and is shown in Figure 3.12.

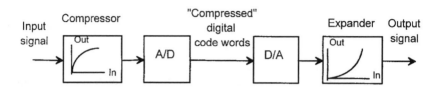

Figure 3.12 Nonuniform quantizing.

One way to understand the companding process is to think of compressing the dynamic range of the analog signal first by compressor circuitry, which amplifies low levels more than higher levels; see Figure 3.12. After this we may use linear quantization and the signal values after compression and linear quantizing are actually nonuniformly quantized. In the decoder of the receiver we use linear quantizing to reproduce the compressed sample values. Then we use a low-pass filter sample sequence to reproduce the compressed analog signal. We then expand this analog signal by amplifying low levels less than high levels in order to cancel out the distortion that was produced by the compressor in the encoder. After linear decoding in the receiver the noise level is the same at any sample level. In expansion a low-level signal is reduced to its original value and quantizing noise is attenuated. This makes the noise level lower at low signal levels than at high signal levels and improves the S/N at low signal levels.

The modern integrated codec (encoder/decoder) chips that are available for PCM coding include both encoder and decoder circuits. They use signal processing technology in order to perform companding and we may not find separate analog nonlinear amplifiers in real-life chips.

An example of a PCM compressor curve for positive analog signal values is presented in Figure 3.13. The horizontal axis represents the original value of an analog voice signal, and the vertical axis gives the output value of the compressor. Uniformly spaced levels of the linear quantizer are shown on the left-hand side. At a certain sampling instant an analog signal value x is quantized according to the curve into one of the quantum levels of $Z(x)$ and this level is then transmitted as a digital word unique to that level.

When a high signal value changes (see change "b" in Figure 3.13), only a couple of quantizing levels are involved. This is adequate because the quantizing noise does not disturb the listener very much if the signal level is high as well. At low levels (see change "a" in Figure 3.13), a small change of signal level uses many quantizing levels; this results in a smaller quantizing error or noise. This improvement of the average S/N at low analog signal levels is essential because noise is most disturbing at low signal levels.

In the decoder, the inverse process is carried out. We could imagine the same curve as in Figure 3.13 but input values are samples at quantum levels

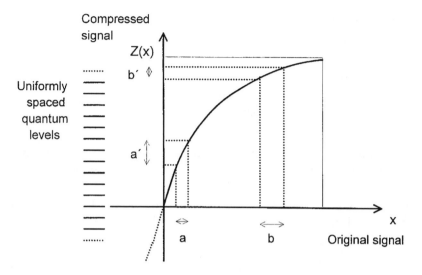

Figure 3.13 Compressor characteristics.

of the vertical axis and the output signal of the expander of the decoder is given as "x" on the horizontal axis. Alternatively we could see an expansion curve as presented in Figure 3.12, where reproduced samples lay on the horizontal axis and the output analog signal is given by the values of the vertical axis according to the curve of the expander.

3.6.5 Companding Algorithms and Performance

As we saw previously, we can improve coding performance if the quantization intervals are not uniform but are allowed to increase with respect to the sample value. If we let quantization intervals be directly proportional to the sample value, the SQR is constant for all signal levels. When the quantization intervals are not uniform (nonlinear quantizing), a nonlinear relationship exists between code words and the sample values that they represent. There are two different nonlinear coding schemes standardized for speech by ITU; they are known as A-law and μ-law coding.

Some key points about these coding schemes are:

- Companding curves are based on the statistics of human voice and many good solutions can be found.

- The two approaches that are standardized internationally are the A-law, which is used in European standard countries (recommendation G.732 of ITU-T), and the μ-law, which is used in North America and Japan (recommendation G.733 of ITU-T).

- These schemes have quite the same quality, but they are not compatible. A conversion device, a transcoder, is needed between countries using different standards.

- Nowadays conversion is a straightforward digital mapping process, where one digital sample value corresponds to another digital value of another coding scheme.

Various compression-expansion characteristics can be chosen to implement the compander. By increasing the amount of compression, we increase the dynamic range at the expense of the S/N for high signal amplitudes. One family of compression characteristics (recommendation G.733) used in North America and Japan is the μ-law companding, which is defined as:

$$Z(x) = \text{sgn}(x) * \frac{(\ln(1 + \mu|x|))}{(\ln(1 + \mu))} \qquad (3.H)$$

where x is the signal value. $Z(x)$ represents the compressed signal, $\text{sgn}(x)$ is the polarity (+ or −) of x, and μ is the constant with a standard value of 255.

Another approach is A-law (recommendation G.732) used in areas of European standard, where the curve is divided into linear and logarithmic sections:

$$Z(x) = \begin{cases} \text{sgn}(x) * \dfrac{(1 + \ln A|x|)}{(1 + \ln A)} & \dfrac{1}{A} < |x| < 1 \\ \dfrac{Ax}{(1 + \ln A)} & -\dfrac{1}{A} < x < \dfrac{1}{A} \end{cases} \qquad (3.I)$$

where x is the signal value, $Z(x)$ represents the compressed signal, $\text{sgn}(x)$ is the polarity of x, and A is a constant with a standard value of 87.6.

In the case of A-law the SQR is constant in the logarithmic section and directly proportional to the signal value in the linear section; see the dashed line curve in Figure 3.14. ITU-T/CCITT recommendations define the continuous curves given by the preceding formulas but approximate them with a curve with linear segments for easier implementation.

3.6.5.1 Companding Performance

As an example of the performance of a nonlinear coding scheme, Figure 3.14 represents the SQR dependence on the signal level for A-law companding. The signal level is measured in dBm0, which we explain at the end of this chapter and may vary within the range of 40 dB, while SQR remains nearly

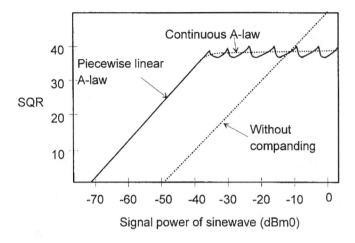

Figure 3.14 Companding performance.

unchanged. However, when the signal level is high, linear quantizing would give better performance as the "without companding" line shows.

We see from Figure 3.14 that at low levels the SQR of A-law companding is more than 20 dB better than linear coding. The curve gives the performance when the signal is a sine wave and the ripple of the curve is a consequence of the approximation of the compression curve with linear segments.

3.6.6 Binary Coding

Finally, in the PCM-encoding process each sample is represented as one in the set of 8-bit binary words. As an example of binary coding, the structure of the 8-bit binary word in the case of European PCM coding, A-law, is defined as follows:

- Bit 1, the *most significant bit* (MSB): The MSB is the first one and it tells the polarity of the sample. Value one represents the positive and zero represents the negative polarity. The sample value zero may create two different code words depending on whether it has a positive or negative polarity.

- Bits 2, 3, and 4: These bits define the segment where the sample value is located. Segments 000 and 001 together form a linear curve for low-level positive or negative samples. Thus, the A-law curve has 13 linear sections.

- Bits 5, 6, 7, and 8: These are the least significant bits and they tell the quantized value of the sample inside one of the segments. Thus each segment is divided in a linear fashion into 16 values.

The structure of the encoded binary word together with the nonlinear relationship between signal values and binary words is shown in Figure 3.15. Note that both the previously described nonlinear compression and linear coding are combined in the same figure.

Finally, after this encoding process, every other bit of the code words is inverted before multiplexing. This inversion was specified to "mix" the digital signal for easier timing of line systems and equipment interfaces. We can, for example, imagine that the signal value stays at a small negative value, which produces the encoded word 00000000, and inversion of every other bit produces the word 01010101. Without the inversion of every other bit we would transmit

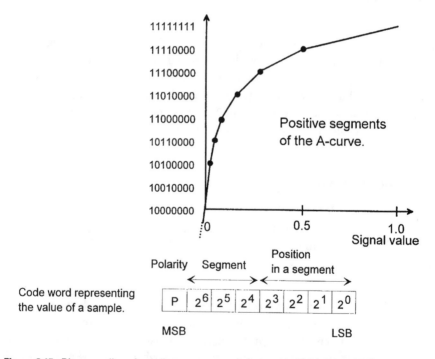

Figure 3.15 Binary coding. A nonlinear compression curve is divided into 13 linear segments. One segment is common to both positive and negative parts of the curve. Bit 1 indicates if the sample is positive or negative. The following three bits define the curve segment and the last four the position in a segment. Before multiplexing every other bit of the word is inverted.

continuous zero and might have difficulties synchronizing the receiver with the received data stream.

3.6.7 PCM Encoder and Decoder

The PCM-coding schemes for digital voice communications were standardized by CCITT (presently ITU-T) at the beginning of the seventies. The standards were based on the technology of those days. The European standard was defined to be slightly different from the American standard, which is why conversion equipment is needed when communicating over the Atlantic sea or from Europe to Japan. Most countries in the world use the European A-law standard. As a conclusion to our discussion about PCM-coding we now look at the block diagrams of the PCM encoder and decoder that contain the processes that we have discussed in previous sections.

3.6.7.1 PCM Encoder

Figure 3.16 presents a block diagram of the PCM encoder according to the European standard. Before actual encoding, the analog signal is filtered into the frequency band from 300 Hz to 3400 Hz. This bandwidth was defined to be acceptable for sufficient quality human voice so that the speaker can be recognized at the other end. This filtering is mandatory to assure that the sampling theorem is satisfied, that is, the analog signal does not contain frequencies higher than half of the sampling frequency. Then the analog signal is sampled at an 8-kHz sampling frequency and the samples are nonlinearly coded into 8-bit words by a quantizer and an encoder.

Words are then converted into serial form and multiplexed with other PCM-coded voice signals into a 2048-kbit/s primary rate signal that contains

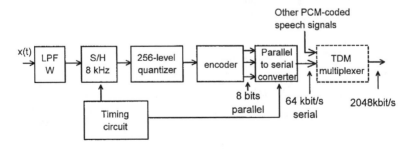

Figure 3.16 PCM encoder. 2048-kbit/s data stream includes 30 PCM speech channels. 64 kbit/s is reserved for signaling and 64 kbit/s for frame alignment. The frame alignment word tells where each channel is located in the data stream. LPF, low-pass filter; S/H, sample and hold citcuit; TDM, time division multiplexer.

30 voice channels according to the European standard. This 2 Mbit/s is a very common data rate in the telecommunications network. For example, digital exchanges build up 2-Mbit/s streams with 30 PCM-coded subscriber interfaces for internal transmission inside the equipment. The multiplexing process is described in Chapter 4.

In the United States the corresponding data rate is 1.544 Mbit/s instead of 2.048 Mbit/s. In this DS1 system each frame contains 24 speech channels and a framing bit. The sampling rate is the same 8 kHz and we get

$$8000 \times \{(8 \times 24) + 1\} = 1.544 \text{ Mbit/s} \tag{3.J}$$

3.6.7.2 PCM Decoder

At the receiver the demultiplexer separates 64-kbit/s individual channels that are then converted into 8-bit parallel sample values, as shown in Figure 3.17. Sample pulses are reconstructed and the resulting series is filtered to create a voice signal that closely resembles the original.

3.7 Other Speech Coding Methods

PCM was standardized during the 1970s, and with the modern signal processing technology many more efficient coding methods have been implemented. By more efficient we mean that we may get better quality at the same data rate or equal quality at a lower data rate. More sophisticated coding schemes are used, for example, in the ISDN, where the ISDN telephone may transmit better quality 7-kHz speech band at 64 kbit/s. Another example that we will

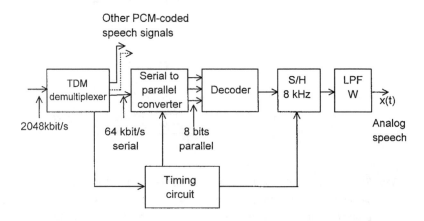

Figure 3.17 PCM decoder. LPF, low-pass filter; S/H, sample and hold circuit; and TDM, time division multiplexer.

briefly review is GSM communications, where speech requires only 13 or 7 kbit/s.

In the following subsection we will review some methods that are used in telecommunications networks in addition to the PCM that we discussed in the previous section.

3.7.1 Adaptive Differential Pulse Code Modulation

In the PCM we encode all sample values independently. We may improve encoding performance by assuming that the next sample value is not independent from the previous one, which is the case in practice.

3.7.1.1 Differential Pulse Code Modulation

In *differential PCM* (DPCM) only the difference between a sample and the previous value is encoded. Because the difference is typically much smaller than the overall value of the sample, we need fewer bits for the same accuracy as in the PCM and the required bit rate is reduced. This coding method is sometimes used for digital video transmission.

3.7.1.2 Delta Modulation

Delta modulation (DM) is a simplified DPCM that transmits binary value "1" if the sample is higher than the previous one. Binary value "0" is transmitted if the signal value has decreased. This simple principle is used sometimes in remote monitoring and control systems to transmit, for example, a measuring result; but it is rarely used for voice communication in telecommunications networks.

3.7.1.3 Adaptive Differential Pulse Code Modulation

An even more efficient method than the DPCM was selected as a standard for voice coding, namely, *adaptive differential PCM* (ADPCM). Further compression is achieved by adapting the predictor and the quantizer to the characteristics of the signal. Both the encoder and the decoder use the same algorithm to estimate the value of the following samples with help of the preceding samples and only the error to this estimate is transmitted; see Figure 3.18. Prediction is based on predefined algorithm. In the original 32-kbit/s ADPCM the difference between predicted and actual sample value is coded with four bits, that is, into fifteen quantum levels, and data rate is half of the conventional PCM. If several subsequent samples vary widely, the quantizing steps are adapted to that change in order to get further improvement of quality. If prediction errors tend to increase, quantizing steps are increased and vice versa.

According to ADPCM standards the commercial voice quality is coded into 32 kbit/s or even into a lower (24 or 16 kbit/s) bit rate. Samples are still

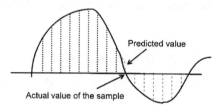

DPCM: The difference between the sample value and the preceding
sample is encoded instead of the full sample value.
ADPCM: The difference between the predicted and actual value of the sample is encoded.
Quantizing steps are increased if the prediction error tends to be high.

Predicted value

Actual value of the sample

Figure 3.18 DPCM and adaptive differential pulse code modulation. The data rate is reduced to at least half for the same voice quality as PCM. Commercial at 32 kbit/s (or 24/16 kbit/s); better quality at 7 kHz bandwidth, voice at 64 kbit/s.

taken at 8 kHz but transmitted with four bits (in the case of 32-kbit/s ADPCM) and the quality is equal (or at least close) to the quality of ordinary PCM.

ADPCM is a relatively recent technique (first standards approved by CCITT in 1984) that is adopted worldwide for digital voice transmission between countries or within a country. It can partly resolve the current compatibility problem between North America's and Europe's PCM formats due to their different companding schemes by acting as a common language between the two PCM schemes.

The original ADPCM recommendation (G.721) was updated in 1988 (Blue Book). The 32-kbit/s coding scheme was updated and some lower bit-rate (24 and 16 kbit/s) options were added.

There is also a recommendation of an ADPCM algorithm (G.722) for coding 7.1-kHz bandwidth audio signals into 64 kbit/s. This coding scheme improves the quality of speech and can be used for good quality voice over ISDN networks.

There are ADPCM systems available on the market that convert two primary rate PCM-streams into one data stream at the same rate using ADPCM. Two ADPCM channels occupy one ordinary PCM channel. These are used by network operators for more efficient use of the long-distance transmission systems. Another application example is PABX networks where offices of private enterprises are interconnected by leased-line 64-kbit/s channels. ADPCM doubles the capacity of these expensive leased lines between PABXs.

The ADPCM coding scheme is based on the statistics of speech and it does not support modem or facsimile signals at higher data rates than 4800 bit/s. Because of this, telecommunication network operators cannot use ADPCM instead of PCM-coding for all calls. This is a problem if ADPCM systems are used inside a telecommunications network. One way to overcome it is to use the ADPCM encoder to detect if there is a data or facsimile connection to be established and, in that case, disable the PCM/ADPCM transcoder for that channel.

Until this point we described *waveform coding* methods only. Waveform coding means that we try to describe the shape or the waveform of the original analog signal—just as PCM, DPCM, and ADPCM do. In a more efficient coding scheme, *source coding,* we consider the characteristics of the voice signal, model it, and send the codes of the models we have used instead of trying to imitate the shape of the signal. An example of the methods that use both of these principles is the voice coding of a cellular network, which we briefly review next.

3.7.2 Speech Coding of GSM

In cellular networks an efficient coding scheme is needed to make maximum use of radio frequencies. The lower the data rate, the narrower frequency band we need, as we will see in Chapter 4. As an example of these efficient coding schemes we now briefly review the principle that is used in the GSM communications.

During the standardization work of the speech coding algorithm for GSM, the goal was to achieve 16-kbit/s data stream with the same speech quality as ordinary PCM. Waveform coding, such as PCM or ADPCM, did not give sufficient quality at this low data rate. Source coding did give a low enough data rate but not good enough quality. Source coding means that the signal is modeled and the codes of the sound elements are sent. In the decoder the speech is reproduced.

A combination of these two basic principles was selected. The maximum processing delay was restricted to be less than or equal to 65 ms, which requires the use of echo cancellers in the network. The original data rate became 13 kbit/s, and it was further reduced to 7 kbit/s in 1995 with a more efficient coding algorithm.

The selected efficient speech coding is always used at the radio path where efficient utilization of transmission channels is more important than in the wireline systems. We cannot make the available frequency band wider, but we can build new optical cable systems when the demand for transmission capacity increases. For switching and interconnecting to a fixed telecommunications network, GSM-coding is changed into ordinary PCM.

3.7.2.1 Principles of Operation

The voice signal is first divided into 20-ms slices. Each slice of the signal is analyzed and the periodicity is noticed. The periodical component is subtracted from the original signal analysis filter and the amplitude of the voice signal is considerably reduced; see Figure 3.19.

The periodical high-power component is transmitted as a set of parameters, and the low-level error or difference signal at the output of the analysis filter is waveform coded. This waveform coding does not require a high bit rate because the amplitude of the error signal is low.

At the receiving end a synthesis filter is used and, with the help of the transmitted coefficients, adds the periodical component to the error signal, which is reproduced from waveform-coded samples.

3.7.3 Summary of Speech Coding Methods

We introduced some important standardized coding methods like PCM, ADPCM, and GSM-radio channel voice coding schemes that are widely used in public telecommunications networks. However, in private PABX networks more efficient coding schemes are sometimes attractive because the charge of leased lines between office sites is based on the chosen bit-rate capacity and we can accept worse quality than in public networks. There are many technologies available, but we will review here only that which gives the lowest data rates, namely, vocoders.

3.7.3.1 Linear Predictive Coders or Vocoders

One way to radically reduce the required bit rate is *linear predictive coding* (LPC). The speech is first synthesized (or modeled) and the resulting parameters are then encoded for transmission instead of the actual signal. This method is also used for speech synthesis (speech generation). These types of algorithms

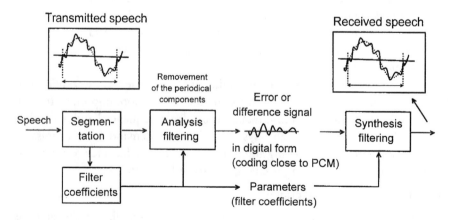

Figure 3.19 The principle of GSM speech coding.

actually try to imitate the human vocal tact, utilizing code books of common phonetic sounds and transferring these codes between the encoder and decoder.

The quality of LPC is worse than the quality of waveform coders. Vocoders sound synthetic. They do not meet the quality requirements of the telephone network, one of which is speaker recognition, but they can be used in private networks. The principle of LPC is also used as a part of the GSM speech coding scheme together with waveform coding, as we saw in the previous section.

3.7.3.2 Quality of Speech Coders

The service quality of a telephone channel is governed by many factors, including volume, distortion, background noise, round trip delay, and echo loss. There are many means to measure the quality. The results in Figure 3.20 are based on a so-called *mean opinion score* (MOS) measurement, where many people have given their opinion about the quality.

The interactive nature of human conversation places a demand on the coder in terms of an acceptable path delay. Subjectively there is noticeable deterioration in perceived channel quality once the delay exceeds 180 ms. Note that via satellite the transmission delay is approximately 250 ms.

Another problem is an echo, which can be noticed if the delay is more than 30 ms to 50 ms. Some echo is always produced at the far end of a 4W/2W hybrid described in Chapter 2 because of the nonideal return loss of the hybrid. This is why, in the case of a long delay (e.g., on satellite channels),

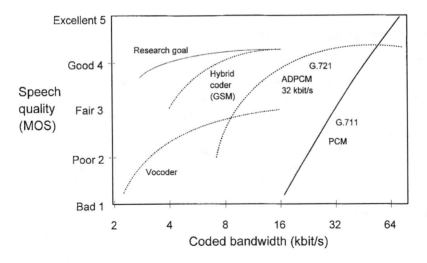

Figure 3.20 Comparison of speech coding techniques.

echo cancellers are needed. The long coding delay (e.g., GSM) also requires echo cancellers.

Figure 3.20 gives a comparison of the coding schemes we discussed in this chapter. The measure of quality is the MOS, which indicates the average opinion by a number of people about the quality of each coding scheme. One example of hybrid coders that utilize both source and waveform coding methods is the GSM coding scheme that was introduced previously.

We introduced some speech coding methods as examples but there are many other standardized speech coding schemes in use at different data rates. Standards of the ITU cover constant bit-rate coders at data rates down to 5 kbit/s. For cellular networks many different lower rate coders are defined in Europe, the United States and Japan and operate at the fixed data rate of 3 to 13 kbit/s. In the United States a variable bit-rate coder is also used in the *code division multiple access* (CDMA) cellular network. It uses a data rate between 1 and 9 kbit/s depending on the speech characteristics.

3.8 Power Levels of Signals and Decibels

In this final section on signals we explain the *decibel* (dB), a measure of signal level. We use this logarithmic measure in the telecommunications network for many purposes, such as to express the voice power, volume measure of voice, as well as transmission and reception power of radio systems such as mobile telephones or an optical line system.

3.8.1 Decibel, Gain and Loss

Along the long-distance communication connection or channel the power of the signal is reduced and amplified over and over again. The signal power needs to be rigidly controlled in order to keep it high enough compared to the background noise and low enough to avoid system overload and resulting distortion.

The reduction of signal strength, attenuation, is expressed in terms of *power loss*. When the signal is regained, this is expressed in terms of *power gain*. Thus an absolute gain of ten corresponds to a loss of 1/10.

3.8.1.1 Decibel

Alexander Graham Bell started to use logarithmic power measures. This was found to be handy and the unit for power gain was named in Bell's honor as a decibel. The gain in decibels is defined to be

$$g_{dB} = 10 \log_{10} g = 10 \lg\left(\frac{P_{out}}{P_{in}}\right) \qquad (3.K)$$

If the output and input powers are the same, the absolute gain and loss both have values of one and the corresponding gain and loss in decibels are 0 dB. If the gain is ten, the corresponding decibel value of gain is 10 dB. The loss is correspondingly 1/10, that is, equal to −10 dB. Thus if the power is reduced, the gain in decibels becomes a negative value. Figure 3.21 presents an element in a telecommunications network with a certain input power and an output power. The formulas of loss (attenuation) and gain are given in the figure as well.

In telecommunication systems we usually have many elements in a chain. If the overall gain or loss needs to be calculated, all gain figures (which often are very large or small numbers) must be multiplied. If the gain of each element is presented in decibels, the figures (that usually have values of less than 100) are added along the chain to determine the overall gain, as shown in Figure 3.21.

Note that the decibel is the measure of power gain and, if we are interested only in voltage levels, impedances must be considered. The voltage and power gains are the same only if the impedances at the points where the power and voltage are measured are the same. The following formula gives the power gain if input and output voltages and impedance are known:

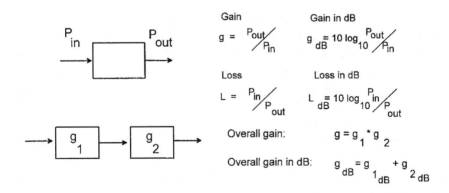

Figure 3.21 Gain, loss and decibels. Gain is often presented in decibels instead of absolute figures because a wide range of values can be represented by small numbers, usually between 0 and 100, and chained systems are easily calculated by adding decibels instead of multiplying by absolute gains. For example, a gain of 100000000 corresponds to a gain of 80 dB.

$$g_{dB} = 10 \lg\left(\frac{P_{out}}{P_{in}}\right) = 20 \lg\left(\frac{V_{out}}{V_{in}}\right) + 10 \lg\left(\frac{Z_{in}}{Z_{out}}\right) \qquad (3.L)$$

The impedances in this equation are assumed to be real numbers.

3.8.2 Power Levels

In the previous section we expressed power ratios in decibels. That does not tell anything about the actual power in Watts, only the ratio. Instead of the actual power in Watts we can use the decibel-based figures also. Power levels in practical systems may vary from nanowatts to tens of Watts, which correspond to a variation from 1 to 1,000,000,000. Power measures based on decibels can be used to express this wide power range in an easy way.

3.8.2.1 Absolute Power Level

The level of absolute power is often expressed in dBm, where the actual power is compared to 1-mW power. The power level in dBm is given by the expression $10 \lg(P/1 \text{ mW})$ dBm. If there is a need to know absolute power in Watts, we can easily calculate it from the given dBm value. Absolute power level dBm is commonly used instead of the absolute power in Watts to express the optical output and received power of optical line systems and the received radio signal strength of a mobile telephone, for example.

3.8.2.2 Relative Power Level

To administer the net-loss of a transmission link, the signal levels of various points in the system are specified in terms of a reference point. This measure is known as dBr. The CCITT recommendations call this point the zero-relative-level point. The reference point may not exist as an accessible point in the network but has long been considered as at the sending terminal of a two-wire switch. It is useful in relating the signal level at one point in the circuit to the signal level at another point.

Note that the relative level is only a figure used to describe how actual signal levels differ from the level at the reference point (or between different points in the network). It does not tell what the actual power is at that point in operational conditions.

If a 0-dBm test tone is applied at the reference point (where relative power level is 0 dBr), the power level at any other point in the circuit is determined directly (in decibels referred to as 1 mW, dBm) as the dBr value of that point, that is, the dBr value tells what the measuring result in dBm should be.

Actual signal power levels or noise power levels are not normally expressed in terms of locally measured values. Instead, powers are expressed in terms of their values at the reference point. For this the measure dBm0 is used. For example, the dBm0 value of noise directly specifies the S/N at that point if the signal has a nominal power level; see the following example.

3.8.2.3 An Example: Power Levels

Let us assume that we have relative levels defined at three points: A, B, and C, as shown in Figure 3.22. We want to determine the following values.

1. The proper signal power in dBm applied to point B in order to determine that A and C are at proper levels;

2. The gain or loss a signal experiences when propagating from point A to point C;

3. Noise level relative to the zero-reference point at B, in dBm0, if −63-dBm noise is measured at point B;

4. The power level of noise that would be measured at point C if −63 dBm is measured at point B and no additional noise is added on the way from points B and C;

5. Zero relative noise level at point C, dBm0, if −63 dBm is measured at point B.

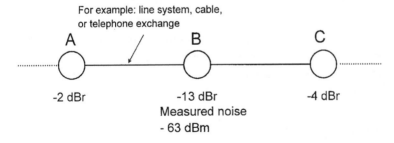

Figure 3.22 An example and some key points about power levels. Nominal relative levels are specified during network engineering and later on the signal level at any point is easily calculated if it is known at one point. Values of relative levels do not tell much about actual operational signal levels. Figures indicate that, for example, at point A a signal is attenuated by 2 dB compared to the network reference point. If a test tone of 0 dB is applied to the network zero-relative-test point, the measurement result should be −2 dBm at A.

Solution

1. Because point B is a −13-dBr relative level point, the proper test tone will be −13 dBm. The zero-relative-level point (= a point where relative power level is 0 dBr) of the network is selected so that 0 dBm is a suitable level for network measurements.

2. The network is designed in such a way that the signal level is dropped from point A to point C by 2 dB, which is the difference between relative levels—the specified dBr values—of points A and C.

3. The question could be expressed another way: What would the noise level be if point B were a 0-dBr level point? The signal and noise should be gained by 13 dB, giving a noise level of −50 dBm, which is the same as −50 dBm0 (in 0-dBr reference point). Another way to understand this is to imagine what would be the difference between the signal, applied at the 0-dBm level to 0-dBr point, and measured noise at point B.

4. The measured noise would be 9 dB stronger at point C than at point B. This is equal to the difference of dBr values, and at point C we would get −54 dBm.

5. The figure is 50 dBm0, which is the same as at point B because at both points the noise is compared to the zero reference point. If this is to be the only noise in the network, 50 dBm0 will be valid at any network point. This shows the advantage of using dBm0. In the case of noise it describes the quality, the noise distance from the nominal signal level regardless of the actual signal level at that point.

We reviewed only some important examples of decibel measures used in the engineering of a telecommunications network. There are many others in use, but we do not consider them here because our objective is to give an introduction to the decibel measure only.

3.8.3 Digital Milliwatt

As we have seen, the PCM systems have a strictly limited operational range. The upper limit is defined by the code word representing the maximum sample value. If an analog signal gets a higher amplitude, it is severely distorted because of clipping. The other limiting factor is quantizing noise, which reduces performance as the signal level decreases.

In a digital international connection, the PCM equipment at both ends must be compatible and convert digital information into the same analog power level and vice versa. Therefore, the control of power level at the PCM encoder input is extremely important. For this purpose ITU-T has defined a digital sequence of code words. By decoding this sequence, a 1-kHz sine wave is produced at the 0-dBm power level; see Figure 3.23. The decoder output level is defined to be the same as the level at the network reference point (0-dBr point) and, thus, can also be written as 0 dBm0. The overload level of the PCM coder is +3.14 dBm0, and at a higher level of 1 kHz the sine wave is distorted.

Measuring systems that generate a bit sequence of the digital milliwatt are available for decoder calibration. When this is done, we can encode 1 kHz and 0 dBm analog signal, loop a digital signal back from the encoder to the decoder and adjust the encoder so that 0 dBm at the input of the encoder produces 0 dBm at the output of the decoder.

Digital milliwatt for European PCM is defined by the data sequence in Table 3.1. Note that before decoding we have to invert every other bit, namely, bits 2, 4, 6, and 8. The PCM decoder produces a 1-kHz sine wave at 0-dBm power level when this sequence is inserted into the digital input of the decoder.

There is often a need to adjust the power level at the interface of PCM equipment so that the following system will not be overloaded. For this purpose there are adjustable amplifiers and attenuators integrated into commercial PCM equipment. Figure 3.23 shows an example where the PCM coder amplifies an

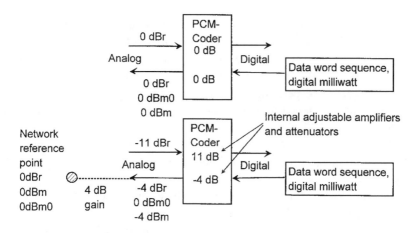

Figure 3.23 Digital milliwatt. The digital milliwatt is a CCITT/ITU-T defined digital signal that produces in D/A conversion 1-kHz sine wave with power of 1 mW (to the reference point). If a PCM-coder is connected to a network point with nonzero relative level, the coder has to be adjusted to attenuate or gain the signal correspondingly.

Table 3.1
Data Sequence for Digital Milliwatt

Word	Bit number							
	1	2	3	4	5	6	7	8
1	0	0	1	1	0	1	0	0
2	0	0	1	0	0	0	0	1
3	0	0	1	0	0	0	0	1
4	0	0	1	1	0	1	0	0
5	1	0	1	1	0	1	0	0
6	1	0	1	0	0	0	0	1
7	1	0	1	0	0	0	0	1
8	1	0	1	1	0	1	0	0
1	0	0	1	1	0	.	.	.
.	0	0	.	.	.	.	.	.

analog input signal by 11 dB before encoding and attenuates the analog signal by 4 dB after decoding.

3.9 Problems and Review Questions

Problem 3.1: What is the wavelength λ of the radio signal: (a) FM-radio, 100 MHz and (b) microwave radio relay system, 10 GHz?

Problem 3.2: A voltage waveform of a signal follows the equation $x(t) = 5 \cos(1*10^3 \ t)$V, where t = time. What are the frequency, amplitude, radian frequency, and periodic time (period) of this signal?

Problem 3.3: Draw the signal $v(t) = 5 \cos(1*10^3 \ t)$V, using vertical scale Volts and horizontal scale milliseconds.

Problem 3.4: Compare digital telecommunications technology with the analog technology and list the most important advantages of the digital technology.

Problem 3.5: What are the main three phases of PCM encoding (A/D-conversion)? Explain how they are performed.

Problem 3.6: What is nonuniform quantizing and why is it used?

Problem 3.7: What is the minimum sampling rate of speech when the frequency band is 300 Hz to 3400 Hz; and what is the minimum sampling frequency for high-fidelity music, 20 Hz to 20 kHz?

Problem 3.8: Draw the spectrum of an analog signal after sampling when the sampling frequency is 8 kHz and the signal that is sampled is a sine wave with a frequency of 1 kHz, 2 kHz, 5 kHz, and 6 kHz. What happens if we want to reconstruct the original signal from the sampled signal with a low-pass filter that has a bandwidth of 4 kHz?

Problem 3.9: The digital *compact disc* (CD) record player is designed for a sound bandwidth of 20 kHz. Linear encoding with 16 bits per each sample is used. Define (a) the minimum sampling rate, (b) the minimum binary data rate per channel (left or right), (c) the maximum SQR, and (d) the average SQR if the average signal level is 30 dB below the maximum value.

Problem 3.10: Estimate what bit rate will be needed for each voice channel in the digital telephone network if linear coding is used. The same performance, SQR at least 40 dB at signal levels higher than −40 dBm0 (sine wave), is required. *Hint*: Look at Figure 3.13 and estimate how much quantizing noise should be reduced at signal level −40 dBm0 and how much longer sample words would be required for this.

Problem 3.11: How long a time of PCM voice or stereo music (assume that CD-quality requires 700 kbit/s for both channels) can be stored in the following media: (a) 1.44 MB (B = Byte = 8 bits) disc and (b) 200-MB memory space of a hard disc?

Problem 3.12: Input power of an amplifier is 2 mW and output power is 1W. What are the power levels (dBm) at the input and output points, and what is the gain of the amplifier in decibels?

Problem 3.13: What is the absolute attenuation L, absolute gain g, attenuation in decibels, gain in decibels, and output power level in dBm of the circuit? Input and output powers are in Table 3.2.

Table 3.2
Input and Output Powers of a Circuit

	P_{in}	P_{out}
(a)	1 mW	1 mW
(b)	1 mW	0.5 mW
(c)	1 mW	4 mW
(d)	10 mW	10 μW
(e)	10 μW	10 mW

Problem 3.14: Figure 3.24 illustrates a telecommunication connection using a geostationary satellite. Calculate the input and output powers of the satellite amplifier and output power of the antenna at the receiving Earth station. Define both power levels in dBm and absolute power in Watts. Use decibels and define power levels, dBm values, first.

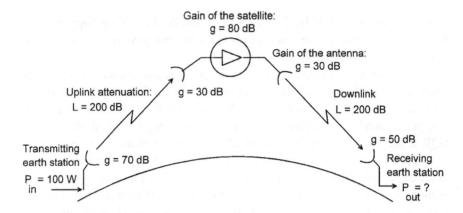

Figure 3.24 Satellite transmission link.

Problem 3.15: The input power of a 40-km cable system is 2W (power at the beginning of the cable). An amplifier with a 64-dB gain is installed at a 24-km distance from the input. Define the signal power level, dBm, and absolute power at the (a) input of the repeater and (b) the output of the system. Attenuation of the cable is 2.5 dB/km.

Problem 3.16: Explain the meaning and purpose of the decibel units dB, dBm, dBr, and dBm0.

Problem 3.17: The sound pressure level is defined as $L_p = 20 \lg p/p_0$ dB, where p denotes sound pressure (in Pascals) and $p_0 = 20\ \mu\text{Pa}$ (20 micropascals, reference level). The threshold for hearing is about 0 dB, and the threshold for pain is about 140 dB. How many times stronger is the sound pressure of the strongest sound that we can hear without pain compared to the weakest one?

Problem 3.18: Explain what the digital milliwatt is.

Problem 3.19: Draw the analog waveform generated by a PCM decoder with a digital milliwatt as the input signal of the decoder. Use Figure 3.15 to determine the approximate analog signal value. What is the periodic time, and what is the frequency of the analog signal produced?

4

Transmission

Transmission is the process of transporting information between end points of a system or network. As we saw in previous chapters, the end-to-end communication distance is often very long and there are many electrical systems on the line. These systems, network elements such as exchanges, are connected to the other elements with connections provided by the transmission systems. In this chapter we discuss the basic restrictions and requirements for transmission and the characteristics of various transmission media and equipment.

4.1 The Basic Concept of a Transmission System

In this first section we look at the basic elements that are present in all transmission systems. We introduce the basic functions of these elements and what their roles for the successful transmission of information are.

4.1.1 The Elements of a Transmission System

The main elements of a communication system are shown in Figure 4.1. The transducers, such as a microphone or a TV camera that we need to convert from an original signal to an electrical form, are omitted; and unwanted disturbances such as electromagnetic interference and noise are included. Note that bidirectional communication requires another system for the simultaneous transmission in the opposite direction.

4.1.1.1 Transmitter

The transmitter processes the input signal and produces a transmitted signal suitable to the characteristics of a transmission channel. The signal processing

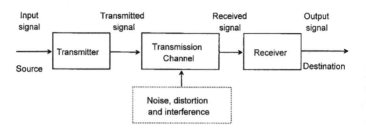

Figure 4.1 Basic concept of transmission system.

for transmission often involves encoding and/or modulation. In the case of optical transmission the conversion from an electrical signal format to an optical signal is carried out in the transmitter.

4.1.1.2 Transmission Channel

The transmission channel is an electrical medium that bridges the distance from the source to the destination. It may be a pair of wires, a coaxial cable, a radio wave, or an optical fiber. Every channel introduces some amount of transmission loss or attenuation and, therefore, the transmitted power progressively decreases with increasing distance. The signal is also distorted in the transmission channel because of different attenuation at different frequencies. We have seen that signals usually contain many frequencies and, if some are attenuated and some are not, the shape of the signal changes. This change is known as distortion. Note that a transmission channel often includes many speech or data channels that are multiplexed into the same cable pair or fiber. The principle of multiplexing is explained later in this chapter.

4.1.1.3 Receiver

The receiver operates on an output signal from the channel in preparation for delivery to the transducer at the destination. Receiver operations include filtering to take away out-of-band noise, amplification to compensate for transmission loss, equalizing to compensate for distortion (different attenuation of frequency components), and demodulation and/or decoding to reverse the signal processing performed at the transmitter. Filtering is another important function at the receiver, for reasons that are discussed later in this chapter.

4.1.1.4 Noise, Distortion, and Interference

Various unwanted factors impact the transmission of a signal. Attenuation is undesirable since it reduces signal strength at the receiver. Even more serious problems are distortion, interference, and noise, the last of which appears as alterations of the signal shape. To decrease the influence of noise, the receiver always includes a filter that passes through only the frequency band of message frequencies and disables the propagation of out-of-band noise.

4.1.2 Signals and Spectra

Electrical communication signals are time-varying quantities such as voltage or current. Although a signal physically exists in the *time domain,* we can also represent it in the *frequency domain* where we view the signal as consisting of sinusoidal components at various frequencies. This frequency domain description is called the *spectrum.*

Any physical signal can be expressed in both domains. In the time domain we draw the amplitude along the time axis and in the frequency domain we draw the amplitude (and phase) along the frequency axis. Although both of them give a perfect description of the signal, both presentations are needed for a better understanding of the phenomena in various systems. The time domain signal is the sum of the spectral sinusoidal components. Fourier analysis gives the mathematical connection between the time and frequency domain descriptions. Here we will only introduce this connection between the time and frequency domain descriptions with a couple of examples. The reader may refer to [1] for the mathematical treatment of the transformation between the time and frequency domains.

In Figure 4.2 the two examples of time domain signals and corresponding spectrums are presented. In the first example we see an ordinary rectangular digital pulse with a duration of T sec and the corresponding spectrum. If, for example, the pulse duration is 1 ms ($T = 1$ ms), the strongest spectral components are below 1 kHz ($1/T = 1/1$ ms $= 1000$ 1/s $= 1$ kHz). From this we get a thumb rule that we can send 1000 pulses of this kind in a second through a channel with a bandwidth of 1 kHz, which corresponds to a 1-kbit/s binary data rate.

In order to increase the data rate we should decrease T and the spectral width and the required bandwidth is increased correspondingly. For example, for a ten times higher data rate we must use ten times shorter pulses, which would require a ten times wider bandwidth.

In another example in Figure 4.2 a digital pulse is sent as a radio frequency burst. Now the spectrum is concentrated around the carrier frequency, f_c, instead of zero frequency. Note that the spectral width still depends only on the pulse duration T in the same way as the previous example. If we now increase the data rate (decrease pulse duration), we make the spectrum wider, which will require a wider radio band. This is one example of digital amplitude modulation known as *amplitude shift keying* (ASK). Other modulation schemes will be discussed later in this chapter.

A pulse and corresponding spectrum

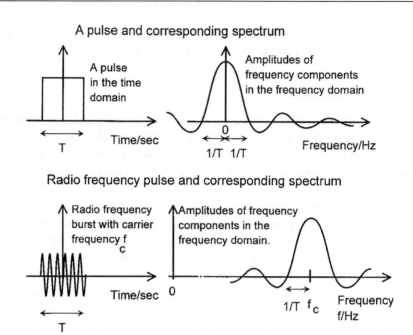

Figure 4.2 Signals in the time domain and the spectrum.

Bandwidth is often one of the restricting factors for transmission. The two preceding examples were aimed to help us understand the connection between the data rate and the required bandwidth. Understanding this we can grasp, for example, why efficient speech coding schemes are required in cellular systems.

4.2 Radio Transmission

In radio transmission we must transfer the spectrum of the message into the radio frequency band for transmission. For this we use *continuous wave* (CW) modulation.

4.2.1 Continuous Wave Modulation Methods

The primary purpose of CW modulation in a communication system is to generate a modulated signal suited to the characteristics of a transmission channel. Modulation is needed in transmission systems to transfer the message spectrum into high radio frequencies. CW modulation is also used in modems where digital data modulates the carrier frequencies inside the voice frequency band.

In CW modulation the message alters the amplitude, frequency, or phase of the high frequency carrier; see Figure 4.3. This alteration is detected in the demodulator of the receiver, and the original message is regenerated.

We saw in Section 3.3 that a cosine wave such as a carrier is defined by three characteristics: amplitude, frequency, and phase. In the CW modulation that is used in radio systems we insert the message into the carrier wave by altering these three factors of the carrier wave.

4.2.2 Amplitude Modulation

The original carrier wave has a constant peak value and a much higher frequency than the modulating signal, the message. When the modulating signal is applied, the peak value of the carrier varies in accordance with the instantaneous value of the modulating signal and the outline wave shape, or envelope, of the modulated wave follows the shape of the original modulating signal, as shown in Figure 4.4. Thus, the unique property of amplitude modulation is that the envelope of the modulated carrier has the same shape as the message.

It can be shown with the help of simple mathematical analysis that when a sinusoidal wave at carrier frequency f_c Hz is amplitude modulated by a sinusoidal modulating signal at message frequency f_m Hz, the modulated wave contains three frequencies:

- The original *carrier frequency*, f_c Hz;
- The *sum* of the carrier and modulating signal frequencies, $(f_c + f_m)$ Hz;
- The *difference* between the carrier and modulating signal frequencies, $(f_c - f_m)$ Hz.

Modulation is used to move the intelligence signal to the frequency band of the transmission channel. This is done by using a high frequency waveform (carrier) to carry the message.

In the modulator the modulating signal (message) systematically alters the carrier wave with the variations in the modulating signal. The message may alter carrier amplitude, frequency or phase.

In the demodulator the variations of the modulated carrier are detected and the original message waveform is reconstructed.

Figure 4.3 Continuous wave modulation.

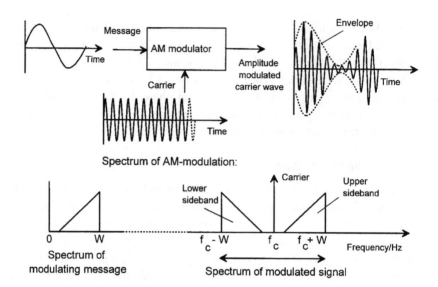

Figure 4.4 Amplitude modulation.

These sum and difference frequencies are new, produced by the *amplitude modulation* (AM) process, and are called *sideband* frequencies.

In this case, the bandwidth of the modulated signal is

$$(f_c + f_m) - (f_c - f_m) = 2f_m \qquad (4.A)$$

If the modulating signal contains multiple frequency components, a band of frequencies such as in speech or music, the AM process transfers the message spectrum with the carrier. The message spectrum appears after the modulation on both sides of the carrier and the required bandwidth is doubled. Figure 4.4 shows an example where the original message with baseband bandwidth W modulates a carrier at the frequency f_c. Each individual frequency that the message contains produces upper and lower side frequencies around the carrier frequency, and complete upper and lower sidebands that contain all the frequencies of the message are obtained.

If the message is in digital format, the amplitude of the carrier is changed rapidly from one value to another. This is called "keying" because in early wireless telegraph systems the carrier was switched on and off with each keystroke by an operator. This type of digital AM is called ASK; its spectrum was presented previously in Figure 4.2.

AM is the oldest modulation method but is still commonly used in modern communication systems together with phase modulation. The original AM has further developed into the *suppressed carrier double-sideband modulation*

(SCDSB), SSB, and VSB versions, which are shortly introduced next. These principles are explained in the frequency domain because some of them are extremely difficult to understand in the time domain; see Figure 4.5.

4.2.2.1 Suppressed Carrier Double-Sideband Modulation

In the case of AM modulation the carrier is in the air even when there is no information to be transmitted. It can be shown that even with the maximum information amplitude, at least 50% of the total power is spent on the carrier wave in AM. In the SCDSB (also called DSB, for short) scheme the carrier wave is suppressed and all the power is used for sidebands that carry the information.

Suppressed Carrier Double-Sideband Modulation, SCDSB

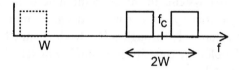

Single-Sideband Modulation, SSB

Vestigial-Sideband Modulation, VSB

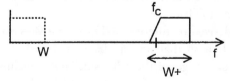

Figure 4.5 Modulation methods. (a) SCDSB, where the carrier is suppressed to save power and the DSB is used as a subprocess in FM-stereo broadcasting. (b) SSB, where only one sideband is transmitted to save bandwidth and the SSB is used in analog carrier systems. (c) VSB, where one sideband and a small fraction of the other are transmitted for improved low-frequency response and the VSB is used in TV-video transmission.

The cost incurred for power saving with the help of SCDSB is having to use more complicated transmitters and receivers, but this is no longer important with the present technology. The detector in the receiver cannot find the message by following the envelope only. The received carrier wave has a phase reversal every time the message crosses zero and, in addition to the amplitude, the phase has to be detected. SCDSB is used, for example, for stereo information processing in present analog FM-radio broadcasting systems and together with phase modulation in digital radio transmission systems.

4.2.2.2   Single-Sideband Modulation

The conventional AM doubles the bandwidth of the message-wasting bandwidth in addition to power. Suppressing one of the sidebands reduces the transmission bandwidth and leads to *single-sideband modulation* (SSB).

The bandwidth of a transmission channel is an especially important restriction of the carrier systems in the telecommunications networks. SSB is used in the analog carrier systems that are designed to transmit as many telephone channels as possible through a bandwidth-limited channel such as a cable. The SSB doubles the capacity (the number of speech channels) compared with AM and SCDSB.

4.2.2.3   Vestigial-Sideband Modulation

Consider a modulating signal, such as the video portion of a television signal, that has a very large bandwidth and significant low-frequency content. The bandwidth conservation principle argues in favor of the SSB, but practical SSB systems have a poor low-frequency response due to the filtering of the other sideband. The SCDSB would be better for this kind of application, but it requires a double bandwidth. Clearly the modulation scheme that makes a compromise between SSB and SCDSB is required, and this is called *vestigial-sideband modulation* (VSB).

The VSB is derived by filtering SCDSB (or AM, VSB is often used with carrier) in such a fashion that one sideband is passed on almost completely while just a trace, or vestige, of the other sideband is included. In the receiver detection circuitry the vestige of the lower sideband is added to the upper sideband. This improves the quality, making the frequency response flat to very low frequencies of the message. This method is used in the present analog TV video transmission.

All the modulation methods described in this section belong to the class of CW modulations that is known as the linear modulation method. Their common properties are:

- The modulated bandwidth never exceeds twice the message spectrum.
- The transmission spectrum is basically the translated message spectrum.
- The destination S/N is never better than if the baseband transmission was used (no modulation at all). This means that the noise power added to the transmitted signal on the line is detected in the receiver together with the wanted modulating signal, and the S/N is not improved in detection.

The exponential modulation methods frequency modulation (FM) and phase modulation (PM) differ on all three counts.

4.2.3 Frequency Modulation

In contrast to linear modulation, exponential modulation is a nonlinear process and, therefore, the modulated spectrum is not related to the message spectrum in a simple fashion.

The modulated waveform after exponential modulation can be expressed by the equation

$$x_c(t) = A_c\cos(\omega_c t + \phi(t)) = A_c\text{Re}[e^{j(\omega_c t + \phi(t))}] \tag{4.A}$$

where $\phi(t)$ represents the varying phase or the frequency containing the message, A_c is the constant amplitude, $\omega_c = 2\pi f_c$ is the angular frequency of the carrier wave, and Re means that we take the real part of the exponential function in brackets.

As we see, the message is inserted into the angle of the carrier wave or in the exponent of the function describing a cosine wave. This is why these modulation methods are called either *angle* or *exponential modulations*.

In FM the instantaneous frequency of the carrier is varied according to the message and its amplitude is kept constant; see Figure 4.6.

Figure 4.6 shows an example where the frequency of the carrier is increased when the value of the modulating message is increased and vice versa. We can assume that FM has good noise performance, because if the amplitude is distorted we can cut it *back* to the constant value in the receiver, thus eliminating most of the external disturbances. In the detector of the receiver only those instants when the signal curve crosses zero voltage need to be detected. The disturbances are highly attenuated because a large amplitude change has only a slight impact on the position of the crossing points. This helps us understand that the noise added to the transmitted signal on the line does not reduce the

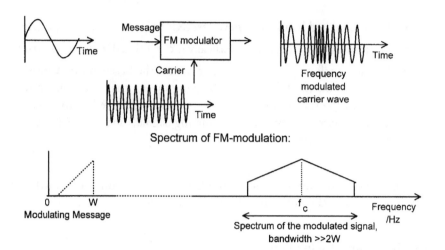

Figure 4.6 Frequency modulation.

post detection S/N as much as in the case of linear modulation. Actually the S/N can be improved in detection.

This advantage is paid for by a wider transmission bandwidth. For example, commercial FM broadcasting uses more than 200 kHz of bandwidth for the transmission of a 15-kHz message band. The characteristics of the spectrum are not as simple as in the case of linear modulation methods and they are beyond the scope of this introduction.

Some older generation voice band modems use the digital form of FM called *frequency shift keying* (FSK). For example, a V.23 modem, 1200 bit/s, uses two frequencies, 1300 Hz for binary "0" and the 2100 Hz for binary "1".

4.2.4 Phase Modulation

PM is another method in the class of exponential modulations. In PM the instantaneous phase, instead of frequency, is varied linearly according to the message. Therefore, if there are discontinuities in the message, there will be discontinuities in the modulated carrier wave as well; see Figure 4.7. The spectral characteristics are nearly the same as in the case of FM.

Figure 4.7 shows an example where the phase of the carrier is increased with the strength of the message. When message returns to zero there is a sudden phase change when the carrier returns to its nominal phase.

In digital binary PM, which is called *phase shift keying* (PSK), the phase of the carrier is varied according to whether the digital signal is 1 or 0. Often we use more than these two phases of the carrier in digital modulation. When

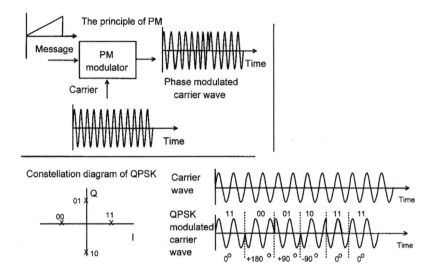

Figure 4.7 Phase modulation and quadrature phase shift keying. The digital form of PM
is called PSK, where carrier phase is 0 or 180 degrees depending on
the binary value of the data. A widely used version of PSK is QPSK, where
two bits at a different time change the carrier to four different phases: 11, 0
degrees phase shift; 01, +90 degrees phase shift; 00, +180 degrees phase
shift; 10, +270 or −90 degrees phase shift.

four carrier phases are used, each phase transmits the value of two binary bits
and we talk about *quadrature phase shift keying* (QPSK). Figure 4.7 illustrates
an example of QPSK. An original and the modulated carrier waves are drawn
in the figure. A pair of bits is taken from the incoming bit stream
(110001101111...) of the modulator simultaneously as the carrier phase is
shifted according to the value of these two bits until the next two bits are
received.

One easily understandable way to describe digital phase modulation is
called a constellation diagram and is shown in Figure 4.7. In the constellation
diagram the I-axis means the in-phase carrier wave and Q stands for the carrier
with a 90-degree phase shift. Each cross in the diagram represents one possible
transmitted "symbol" or waveform that represents binary values of two bits
in the example of Figure 4.7. We can see from the figure that the bit combination
01 is sent as a carrier with a +90-degree phase shift. The distance of the cross
from the origin of the diagram tells the carrier amplitude that it is the same
for all symbols in our example in Figure 4.7.

If we use two amplitude values in addition to the four phases we will
have four more crosses in the constellation diagram. Each of the crosses will

represent the values of three subsequent bits. This combination of phase and amplitude modulations is called *quadrature amplitude modulation* (QAM).

Phase modulation (together with amplitude modulation) is used in many digital transmission systems, such as in digital radio relay systems, voice band modems, and some modern digital cellular telephone systems.

4.2.5 Allocation of the Electromagnetic Spectrum

The signal transmission over an appreciable distance always involves the traveling of an electromagnetic wave, with or without a guiding medium. The efficiency of any particular transmission method depends upon the frequency of the signal being transmitted. With the help of carrier wave modulation the spectrum of the message is transferred to the suitable frequency band of the medium.

The usage of frequency bands is controlled internationally by ITU-R and nationally by national telecommunications authorities. Radio systems are often the most economical solutions when new connections are required and there are no free cables between the end points of the connection. Figure 4.8 illustrates the frequency range that is used in telecommunications and shows some examples of the usage of different frequencies.

In Figure 4.8, according to [1], the electromagnetic spectrum used in telecommunications is shown together with typical transmission media, the propagation modes, and some application examples.

However, there is one important problem with the radio systems that restricts the usage of radio communications, namely, the lack of frequency bands. The most suitable bands are overcrowded, and new technical inventions are needed in order to overcome this problem. Among these are, for example, cellular mobile systems with small cell areas that enable them to use frequencies again in other cells of the same network, narrow-beam radio relay systems, sophisticated modulation schemes for radio relays, and modern digital broadcasting systems. We saw in Section 4.2.4 that we can decrease the modulation rate, and correspondingly the required bandwidth, with the help of more complicated modulation schemes.

4.2.5.1 Wavelength and Frequency

The wavelength, shown on the left-hand side of Figure 4.8, tells the propagation distance during one cycle of the radio wave. It is related to the frequency and speed of light according to $\lambda = c/f$, where λ is the wavelength in meters; c is the propagation speed of light in meters per second, 300,000 km/s; and f is the frequency in Hz = 1/sec.

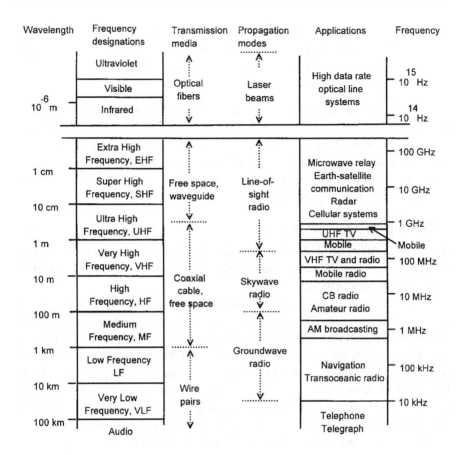

Figure 4.8 The electromagnetic spectrum.

4.2.5.2 Propagation Modes

Radio waves at different frequency bands propagate in different propagation modes. They are very briefly explained as follows.

- *Ground wave*: The radio wave follows the surface of the Earth, and thus communication over the horizon is possible.

- *Skywave*: The radio wave is reflected from the ionosphere back to Earth. The wave is reflected back from the Earth's surface and back to the Earth again making long-distance communication possible. The communication quality is not stable because the characteristics of the ionosphere vary with time.

- *Line-of-sight*: The radio wave propagates along the straight line from the transmitter to the receiver. A general requirement for good performance is that the receiving antenna is visible from the transmitter. The radio frequencies above 100 MHz that propagate in line-of-sight mode are used in most modern communication systems.

As the demand for radio communications has increased, higher and higher frequencies have been put into use. However, as we will see in the next section, the attenuation of the radio wave increases with frequency and at extremely high frequencies, approximately beyond 100 GHz. Even weather conditions affect attenuation. This is why there are no applications at frequencies higher than the *extra high frequency* (EHF) band; see Figure 4.8.

4.2.5.3 Optical Communications

At the frequencies of visible light a controlled transmission medium, optical fiber, has come into widespread use. It provides very low attenuation especially at the frequencies of infrared light.

The commercial optical communication systems of today use binary light pulses for transmission. The transmitted information is usually in binary form, which means that the receiver either detects light or does not. The present optical systems in use do not utilize the frequency content of the transmitted light at all. The whole bandwidth of the fiber is usually occupied by a single binary signal. If we could use light at a certain frequency as a carrier wave, we could increase capacity dramatically utilizing the CW modulation methods discussed previously. The utilization of this so-called coherent optical technology will make huge transmission capacities available in the future.

4.2.6   Attenuation of a Radio Wave, Free Space Loss

The radio wave above 100 MHz travels a direct path from the transmitting antenna to the receiving antenna. This propagation mode is called line-of-sight propagation.

The power of the radio wave reduces with distance just as a cable attenuates propagating electrical signal. The attenuation of a radio wave, free space loss on the line-of-sight path is due to the spherical dispersion of the radio wave. The transmitted power is distributed over a ball surface and the power per unit area decreases in proportion to the square of the radius because the area of the ball surface increases in proportion to the square of the radius. The area of the ball surface follows $A = 4\pi l^2$, where l is the radius.

The receiving antenna is able to receive the power that passes through its capture area. The capture area of the receiving antenna is proportional to the square of the wavelength, $A_e = \lambda^2/(4\pi)$. From these two facts we can easily derive that the free space loss, that is, the ratio of transmitted power and the received power, is

$$L = \left(\frac{(4\pi l)}{\lambda}\right)^2 = \left(\frac{(4\pi fl)}{c}\right)^2 \tag{4.B}$$

where λ is the wavelength, f is the frequency of the signal, c is the speed of light, and l is the transmission distance; see Figure 4.9.

We usually prefer to describe attenuation or loss in decibels instead of the absolute value given by the previous equations. We get the formula that gives dB values by taking $L_{dB} = 10 \log_{10}L$. Now, if we express the frequency in GHz and distance in kilometers, we get the freespace attenuation of a radio wave in decibels as

$$L_{dB} = 92.4 + 20 \log_{10} f_{GHz} + 20 \log_{10} l_{km} \tag{4.D}$$

We see that the loss or attenuation is proportional to 20 times the logarithm of frequency and distance. So if the distance or frequency is doubled, the attenuation increases by 6 dB. If we want to maintain the received power, we have to increase the transmitted power by 6 dB, which means four times higher transmission power. This comes from the fact that the power ratio in decibels is $10 \log_{10}(P_n/P_0)$dB, as we saw in Chapter 3.

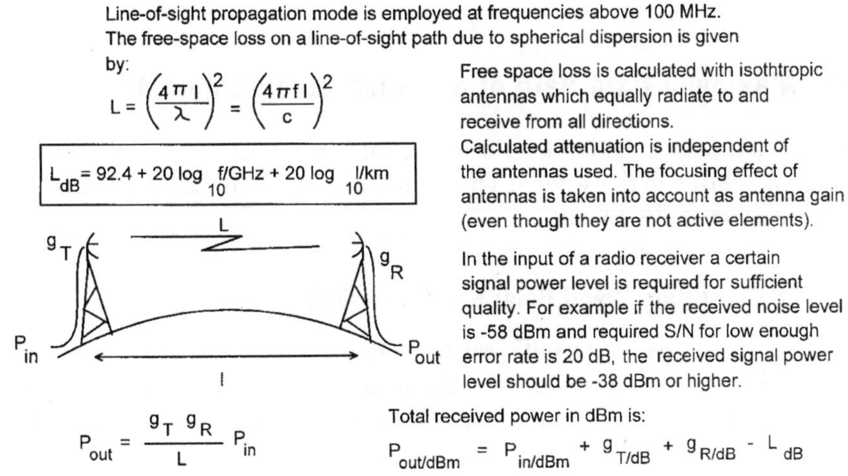

Line-of-sight propagation mode is employed at frequencies above 100 MHz. The free-space loss on a line-of-sight path due to spherical dispersion is given by:

$$L = \left(\frac{4\pi l}{\lambda}\right)^2 = \left(\frac{4\pi fl}{c}\right)^2$$

$$L_{dB} = 92.4 + 20 \log_{10} f/GHz + 20 \log_{10} l/km$$

Free space loss is calculated with isothropic antennas which equally radiate to and receive from all directions. Calculated attenuation is independent of the antennas used. The focusing effect of antennas is taken into account as antenna gain (even though they are not active elements).

In the input of a radio receiver a certain signal power level is required for sufficient quality. For example if the received noise level is -58 dBm and required S/N for low enough error rate is 20 dB, the received signal power level should be -38 dBm or higher.

$$P_{out} = \frac{g_T \, g_R}{L} P_{in}$$

Total received power in dBm is:

$$P_{out/dBm} = P_{in/dBm} + g_{T/dB} + g_{R/dB} - L_{dB}$$

Figure 4.9 Attenuation of the radio wave.

The free-space loss shown in Figure 4.9 gives optimistic results in actual conditions. Additional attenuation is introduced if there is a hill or a building on or close to the straight line between the transmitting and receiving antennas.

4.2.6.1 Antennas

Link loss was calculated assuming that antennas are isotropic, which means that they transmit and receive equally to and from all directions. This assumption keeps the attenuation independent of the antennas in use. However, practical antennas have a focusing effect that acts like amplification, compensating for some of the propagation loss. This focusing effect can be expressed as a gain of an antenna, although a passive antenna cannot actually amplify the signal. The maximum transmitting and receiving gain of an antenna with effective aperture area A_e is [1]

$$g = \frac{(4\pi A_e)}{\lambda^2} = \frac{(4\pi A_e f^2)}{c^2} \qquad (4.E)$$

The value of A_e for a dish or horn antenna approximately equals its physical area, and large parabolic dishes may provide gains in excess of 60 dB.

In this section we reviewed radio transmission at different frequencies and modulation methods that are used to transfer a message to the radio frequency band for transmission. We also examined the propagation loss of the radio wave. In the following section we will look at the characteristics of transmission channels and how the maximum transmission data rate depends on the bandwidth and noise of the channel.

4.3 Maximum Data Rate of a Transmission Channel

There is a fundamental limit for data rate through any transmission channels, as we will see later in this section. The main restricting factors are the bandwidth and the noise of the channel.

4.3.1 Symbol Rate (Baud Rate) and Bandwidth

Communication requires a sufficient transmission bandwidth to accommodate the signal spectrum, otherwise severe distortion will result. For example, several megahertz are needed for an analog television video signal, while the much slower variations of a voice signal fit into a 4-kHz frequency band.

Every communication channel has a finite bandwidth. The higher data rate to be transmitted, the shorter digital pulses used, as we saw in Section

4.1. The shorter the pulses used for transmission, the wider the bandwidth required, as we saw in Figure 4.2. When a signal changes rapidly in time, its frequency content or spectrum extends over a wide frequency range and we say that the signal has a wide bandwidth.

Figure 4.10 shows the shape of a rectangular pulse with duration T before and after it passed through an ideal low-pass channel of bandwidth B. For example, if the duration of the pulse is 1 ms, distorted pulses are shown in the figure for the channel with bandwidths $B = 2*1/T = 2$ kHz, $B = 1/T = 1$ kHz, $B = 1/2*1/T = 500$ Hz, and $B = 250$ Hz. If the next pulse is sent immediately after the one in the figure, the detection of the pulse value will be impossible if the bandwidth is too narrow. The spread of pulses over other pulses is called *intersymbol interference.*

In baseband transmission a digital signal with r symbols per second, Bauds, requires the transmission bandwidth B in Hertz: $B \geq r/2$. Thus, the available bandwidth in Hertz determines the maximum symbol rate in Bauds. Note that the symbol is not necessarily the same as the bit, but it can carry a set of bits if it is allowed to get many different values.

We can find the theoretical maximum of the symbol or Baud rate with the help of a special pulse called the sinc-pulse. The shape of the sinc-pulse is drawn in Figure 4.10, and it has zero crossings at regular intervals $1/(2W)$. With the help of Fourier analysis we can show that this kind of pulse has no spectral components at frequencies higher than W. If the channel is an ideal low-pass channel with a bandwidth higher than W, it is suitable for transmitting

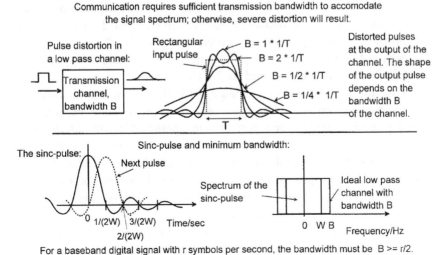

Figure 4.10 Pulse distortion and bandwidth.

sinc-pulses that have their first zero point at $t = 1/(2W)$ without distortion. The shape of the pulse remains the same because all frequency components are the same at the output as at the input of the channel.

The sinc-pulses have zero crossings at regular periods in time. These periods are $1/(2W)$ sec for a sinc-pulse with a spectrum up to frequency W, as shown in Figure 4.10. We can transmit the next pulse at the time instant $1/(2W)$ so that the previous pulse has no influence on the reception because it crosses zero at that time instant. The decision for the value of the pulse is made in the receiver exactly at time points $n*1/(2W)$, where $n = \pm1, \pm2, \pm3, \ldots$; the duration between pulses becomes $T = 1/(2W)$; and the corresponding data rate $r = 1/T = 2W$. If we increase the data rate now so that $W = B$, we get the theoretical maximum and can say that the symbol rate and bandwidth are related according to $r \leq 2B$ and $B \geq r/2$.

This kind of pulse does not exist in reality, but the result gives the theoretical maximum symbol rate through a low-pass channel. In real life systems quite similar pulse shapes are in use and typically a 1.5 to 2 times wider bandwidth is needed.

4.3.2 Symbol Rate and Bit Rate

In digital communications a set of discrete symbols is employed. Binary systems have only two values that represent binary digits 1 and 0. In the previous section we found that the fundamental limit of the symbol rate is twice the bandwidth of the channel. With the help of the symbols with multiple values, the data rate, bit/s, can be increased. As an example, with four pulse values we could transmit the equivalent of two bit binary words 00, 01, 10, and 11. Thus each pulse would carry the information of two bits. So 1 symbol per second (1 Baud) would correspond to 2 bit/s.

If we use a sinc-pulse as in Figure 4.10, the preceding and following pulses do not influence the detection of a transmitted pulse because each received pulse is measured at a zero crossing point $n*1/(2W)$ of the other pulses. We may increase the number of peak values of sinc-pulses from 2 to 4 and from 4 to 8, for example, in order to increase the bit rate while keeping the symbol rate unchanged. In Figure 4.11 we drew the symbols as rectangular pulses for a simple example in which we used four symbol values and each symbol carries two bits of information.

The unit of symbol rate, sometimes called the modulation rate, is Baud (symbols per second). Note that the transmission rate in Bauds may represent a much higher transmission rate in bit/s. An equation in Figure 4.11 gives the dependence of bit rate on modulation or symbol rate and the number of

Symbol rate is sometimes called modulation rate. It tells how many times in a second signal characteristics may change on the line.

Symbol rate is given in Bauds and it is not the same as bit rate, bits/sec. We can for example code 2 bits into one symbol of four values, or 8 bits into a symbol of 256 values. By transmitting one symbol, 2 or 8 bits are transmitted and 1 Baud corresponds to 2 or 8 bits/sec.

An example:
If four-level symbols
are used, each of them
transmits two bits.

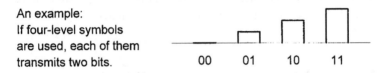

00   01   10   11

Symbol rate is measured in Baud (= symbols/sec). Only in the case of binary signaling (one symbol = one bit), Baud = Bit/sec.

Binary data rate:

$$C = r * \log_2 M$$

C = Binary data (information) rate, bits/s
r = Symbol rate, Bauds
M = Number of symbol values

Figure 4.11 Symbol rate and bit rate.

symbol values. Table 4.1 shows how the bit rate of a system depends on the number of symbol values.

In the preceding examples the amplitudes of the pulses contain the information. This principle is known as PAM. This is not the only alternative, since we can also use other characteristics of the signal to create multiple symbol

Table 4.1
Bit Rate of a System Using Multiple Symbol Values

Number of Symbol Values	Bit Rate Compared with the Symbol Rate
2	the same as symbol rate
4	2 times symbol rate
8	3 times symbol rate
16	4 times symbol rate
32	5 times symbol rate
.	.
256	8 times symbol rate
.	.
etc.	etc.

values, such as the phase of a carrier, as we did in the case of QPSK in Figure 4.7. There we used a certain modulation rate in Bauds (how many times the phase can change in a second) and transmitted bits at a double rate. Using another carrier amplitude value in addition to the four phases we could have eight symbol values, as we discussed in Section 4.2.4, and the bit rate would be three times the modulation rate.

As we see in Table 4.1, by increasing the number of different symbols used in the system, the data rate could be increased without a limit if there were no other limitations than bandwidth. This is not possible in practice because of the noise. The influence of noise is discussed in the following subsections.

4.3.3 Maximum Capacity of a Transmission Channel

We saw previously that the bandwidth of a channel sets the limit to the symbol rate in Bauds but not to the information data rate. In 1948, Claude Shannon published a study of the theoretical maximum data rate in the case of a channel subject to random (thermal) noise.

We measure a noise relative to a signal in terms of the S/N. Noise degrades fidelity in analog communication and produces errors in digital communication. The S/N is usually expressed in decibels as

$$S/N_{dB} = 10 \log_{10}(S/N) \text{ dB} \qquad (4.F)$$

Taking both bandwidth and noise into account, Shannon stated that the error-free bit rate through any transmission channel cannot exceed the maximum capacity C of the channel given by

$$C = B \log_2(1 + S/N) \qquad (4.G)$$

where C is the maximum information data rate (bit/s); B is the bandwidth (Hz); S, the signal power; N, the noise power; and S/N, the signal-to-noise power ratio (absolute power ratio, not in decibels).

The formula gives a theoretical limit for the data rate with an arbitrary low error rate when an arbitrary long error correction code is used. It also assumes that the signal has a Gaussian distribution as the noise does, which is not the case in practice. The influence of noise in the case of binary and multiple value signaling is explained in Figure 4.12.

The signal power and, thus, the highest value of the signal are always restricted to a certain maximum value. Then, the more symbol values we use,

Bandwidth gives the limit to the symbol rate, but using symbols with multiple values we could increase information rate without limit, if this were the only limiting factor.

All systems and channels have a certain limit for the maximum symbol value (for example pulse amplitude) that can be used. Thus, the more symbol values are used, the closer they are to each other and the weaker noise can cause errors, i.e. in the receiving end a symbol is detected as a different symbol.

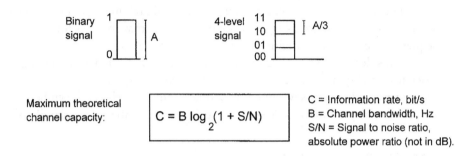

Figure 4.12 The maximum capacity of a transmission channel.

the closer they are to each other and the lower noise level can cause errors. Thus, a higher bit rate requires a wider bandwidth that allows a higher symbol rate. Alternatively, a better S/N is required to allow for more symbol values.

The example in Figure 4.12 shows what happens to the distance between symbol values when the maximum amplitude is A and four symbol values are used instead of binary symbols that have only two values. In our examples we have used symbols with different amplitudes. This transmission scheme is called PAM. Transmission of this kind of pulses without modulation is called baseband transmission.

In radio systems or modems that use continuous-wave modulation, different phases of a carrier wave are often used in addition to the multiple amplitudes. If we use four phases and two amplitudes, each symbol has eight possible values and each symbol carries three bits. This transmission method is called *quadrature amplitude modulation* (QAM). The constellation diagram in Figure 4.7 would in this case have one additional signal point between each signal point and the origin. However, modulation moves the spectrum of the pulse from low frequencies to carrier frequencies and bandwidth is typically doubled when compared with baseband systems. This is why the symbol rate in radio systems is less than or equal to the transmission bandwidth, that is, $r \leq B_T$, where r is the symbol rate in Bauds and B_T is the transmission bandwidth in Hertz.

The accurate requirement of bandwidth depends on the modulation scheme in use; its study is beyond the scope of this book.

4.4 Coding

We described modulation as the processing of a signal for efficient transmission in a different frequency band than where the information originally exists. Coding is a digital symbol-processing operation in which the digital form of the information is changed for improved communication.

The operation of *encoding* transforms a digital message into a new sequence of symbols. *Decoding* is the opposite process that converts the encoded sequence back into the original message; see Figure 4.13.

Consider a computer terminal with a keypad. Each key represents a discrete digital symbol. Uncoded transmission would require as many different waveforms as there are keys, one for each key. Alternatively, each symbol can be presented by a binary codeword consisting of a number of binary digits.

4.4.1 The Purpose of Coding

One purpose of coding is to make the form of the spectrum of a digital signal suitable for a certain communication media. This kind of coding is called line coding. These line codes usually have no DC-content (direct current, frequency component at 0 Hz). We want to get rid of the DC that does not transmit any information but wastes power.

Another reason for line encoding is to help synchronize the receiver. In digital transmission the receiver must be synchronized with the transmitter in order to receive the information when each new symbol arrives. For this the data should be transmitted in the form that contains synchronization information so that there is no need to transmit an additional clock or timing signal.

The systems that use only line coding, but not modulation, are called baseband transmission systems. The spectrum of the line signal is still in the

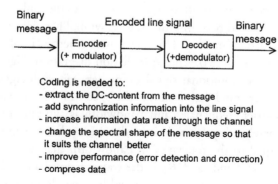

Binary message

Encoded line signal

Binary message

Encoder (+ modulator)

Decoder (+demodulator)

Coding is needed to:
- extract the DC-content from the message
- add synchronization information into the line signal
- increase information data rate through the channel
- change the spectral shape of the message so that it suits the channel better
- improve performance (error detection and correction)
- compress data

Figure 4.13 Coding.

frequency range of the original message, "base band." In radio systems both coding and modulation are used.

In the line coding example in Figure 4.14, each sequence of two data bits is encoded into four-level pulses for transmission. At the receiving end decoding is carried out and the original bits are regenerated. Note that the symbol rate on the line is half of the bit rate seen by the data source and the destination.

We often combine coding and modulation, and instead of four or more pulse amplitude values we transmit four or more different phases. This is called QPSK. QPSK can be seen as a combination of four-level line coding followed by ordinary phase modulation.

We use coding for many other purposes as well, as listed in Figure 4.13, for example for error handling. By appending extra check digits to each binary word, we can detect or even correct errors that occur on the line. Error-control coding increases both bandwidth and hardware complexity, but it pays off in terms of nearly error-free digital communication even when the S/N is low.

Still another purpose for coding is compressing information. Using data compression we may reduce the disc space needed to store data in a computer. In the same way we can decrease the data rate on the line to a small fraction of the original information data rate. We could, for example, use very short codes for the most common characters instead of the full seven-bit ASCII code. Rarely needed characters would use long codes and the total data rate would be reduced. The study of compression methods is a complex matter and will not be covered in any further detail here.

Each two-bit sequence from the source is encoded into one four-level pulse to the line. The sole purpose for coding in this example is to cut the required bandwidth to half.

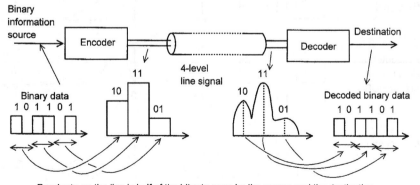

Baud rate on the line is half of the bit rate seen by the source and the destination.

Figure 4.14 An example of the line coding.

4.4.2 Spectrum of Common Line Codes

In order to see what kind of impact encoding has on the spectrum, we will look at the characteristics of some common line codes. Figure 4.15 presents their power density spectrums. That shows how the signal power of random data is distributed over the frequencies.

4.4.2.1 Non-return to Zero

Non-return to zero (NRZ) is the most common form of digital signal that is used internally in digital systems. Each symbol has a constant value corresponding to binary symbol values 1 and 0. The spectrum has a high DC component, and there are no discrete spectral components at the harmonic frequencies of the data rate. The harmonic frequencies are multiples of the data rate. An external clock is always needed for the timing of the receiver.

4.4.2.2 Return to Zero

In the case of *return to zero* (RZ) each symbol is cut to two parts. The first half of the symbol represents the binary value and the second half of the symbol is always set to zero. Because pulses are shorter than in the case of NRZ, the spectrum is wider, as we saw in Figure 4.2, and the spectrum has strong discrete components at the harmonic frequencies of the data rate. With the help of these components, the timing information can be extracted from the signal

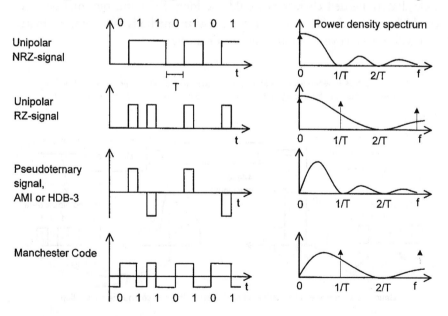

Figure 4.15 Common line codes and their power spectrums.

spectrum and an external clock is not necessarily needed. However, because RZ code has high low-frequency content and a wide spectrum (see Figure 4.15), it is never used in long-distance transmission. One problem with it also is that synchronization is lost if the data content is continuously zero.

4.4.2.3 Alternate Mark Inversion

If every other mark or "1" of the NRZ or RZ-symbols is transmitted as an inverted voltage polarity, an *alternate mark inversion* (AMI) signal is produced. The advantage of this is that no DC component is present on the transmission line. The DC component is unwanted because it does not carry any information but wastes power. With the help of this kind of code we can avoid the problem caused by transformers on the line. Transformers are needed on the copper cable line for matching impedance, for overvoltage or surge protection, and other purposes. Direct current does not propagate through transformers.

AMI code is used in the American telecommunications network in primary rate 1.5-Mbit/s transmission systems. We may extract the timing information by rectifying the AMI signal into an RZ signal in the receiver and then the discrete spectral components appear as in the spectrum of RZ-code in Figure 4.15.

4.4.2.4 High-Density Bipolar 3

High-density bipolar 3 (HDB-3) is developed from AMI and standardized for European primary rate 2-Mbit/s systems. HDB-3 overcomes the problem of the original AMI-code that occurs in the timing when there is a data message that contains long periods of subsequent zeros. In this coding scheme, a pulse with the same polarity as the previous one is added in such a way that no more than three sequential zeros are allowed. In the decoder these pulses are taken away according to the AMI-coding rule that they violate.

4.4.2.5 Manchester Coding

Manchester coding is used in LANs. Binary digit "1" is coded as a "+ to −" transition and binary "0" as a "− to +" transition. The most important advantage of the Manchester code is that each symbol contains the timing information and the receiver needs only to detect the transition in the middle of each received symbol in order to extract the clock signal. Its main disadvantage is a wide spectrum because of short pulses, and this is why it is suitable for LANs but not for long-distance transmission.

4.5 Regeneration

In long-haul transmission the transmitted signal is attenuated and amplifiers or repeaters are needed. *Analog amplifiers* amplify the signal at the input and contain both the desired message and channel noise. In every amplifier and cable section some noise is added and the S/N decreases on the line.

Unlike analog amplifiers, digital repeaters are regenerative. A regenerative repeater station consists of an equalizing amplifier that compensates the distortion and filters out the out-of-band noise and a comparator; see Figure 4.16. Output of the comparator is high if the input signal is above the threshold voltage V_{ref} and low if the input is below the threshold value. The regenerator also contains timing circuitry, which extracts the clock signal from the received data, and a D-type flip-flop, which decides if a digit is high ("1") or low ("0") at the instant of the rising edge of the clock signal; see Figure 4.16. At the rising edge of the clock signal the input value is read into the output by the D-type flip-flop. The output value remains the same until the next rising edge of the clock signal. The operation principle of a binary regenerative repeater is presented in Figure 4.16. The regenerated digits that contain no noise are delivered to the destination or via a cable to the next repeater station (in the case of an intermediate repeater).

If the equalized signal is below threshold V_{ref} at the input of the comparator, the output is low and a zero is regenerated at the rising edge of the clock signal. If noise is too high, the input of the comparator may be above threshold even though a zero is transmitted. If this occurs at the rising edge of the clock signal, the value "1" is regenerated and an error has occurred. In the same way high values may be in error if noise reduces the high amplitude value below the threshold level at an instant of the rising edge of the clock signal. Then "0" is regenerated and an error has occurred.

How frequently errors occur depends on the noise level, that is, on the S/N. If noise is assumed to have a *Gaussian amplitude distribution* (as thermal

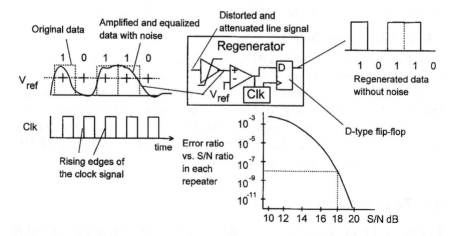

Figure 4.16 Operation principle of a regenerative repeater.

noise does), the error rate (bit error probability) follows the shape of the curve, error ratio vs. S/N, in Figure 4.16.

As an example let us assume that we have a channel, for instance a cable, that attenuates signal so much that the resulting S/N in the repeater is 15 dB. Error rate would then be around $1*10^{-5}$ according to the curve in Figure 4.16. If we place a new repeater in the middle of the repeater section (in the middle of the cable), attenuation of the signal is 3 dB less, giving an S/N value of 18 dB in both repeaters, and the error rate at both repeaters would be $1*10^{-8}$. This means that one error occurs on average after 100,000,000 correct bits. Now we have two repeaters and an overall error rate of $2*10^{-8}$ because each of them creates on average one error in each sequence of $1*10^8$ bits. We can see that the improvement of 3 dB in the S/N that we achieved with the help of the new repeater reduces the number of errors by a factor of 0.001.

In practice, the error rate is even better. We get close to error-free transmission and the exact equivalent of the original signal is received in the end regardless of the distance (the number of repeaters).

The error rate decreases rapidly with noise because of the Gaussian nature of thermal noise. Not only thermal but many other types of noise in real-life systems are assumed to follow a Gaussian distribution. With this model the reduction of noise by 1 dB improves the error rate by a factor of ten or more, as seen in Figure 4.16. The digital transmission systems installed in telecommunications networks are designed in such a way that noise is low enough in all the regenerators and the error rate is extremely low. For example, optical line systems usually have a design practice of a worst case life-time error rate of $1*10^{-10}$. In normal operational conditions the error rate is several orders of magnitude even better and they operate nearly error-free.

From the error rate curve in Figure 4.16 we see how the error rate depends on the S/N. From the error rate we can easily calculate the mean time between errors when the data rate is known. Table 4.2 gives examples of error rates and mean times between errors for a 64-kbit/s (ISDN B-channel) data channel.

Digital systems have a certain threshold value for a S/N. From the curve in Figure 4.16 and from Table 4.2 we find that if S/N is worse than 18 dB, errors occur frequently. At a few decibels better value of S/N the transmission is almost error-free. The S/N values in the curve and in Table 4.2 are examples and are based on certain assumptions. The actual S/N value in decibels at a certain error rate of a specific system depends on the system characteristics and how the S/N is defined and measured. However, the shape of the error curve is the same as in Figure 4.16 and the threshold value is usually between 8 dB and 20 dB.

When the S/N of a digital system decreases, errors occur more and more frequently; and when the error rate becomes too high, information is lost. The

Table 4.2
Examples of Error Rates and Mean Times Between Errors for a 64-kbit/s Channel

S/N dB	Error Rate	Mean Time Between Errors
10.3	10^{-2}	1.5 ms
14.4	10^{-4}	150 ms
16.6	10^{-6}	15 sec
18	10^{-8}	26 minutes
19	10^{-10}	2 days
20	10^{-12}	6 months
21	10^{-14}	50 years

error rate $1*10^{-3}$ is standardized to be the worst allowed communication quality for speech in the telecommunications network. If the error rate becomes worse, on-going calls are cut off.

4.6 Multiplexing

Multiplexing is a process that combines several signals for simultaneous transmission on one transmission channel. Most of the transmission systems in the telecommunications network contain more capacity than is required by a single user. It is economically feasible to utilize the available bandwidth of optical fiber or coaxial cable or a radio system by a single high-capacity system that multiple users share. The main principles of multiplexing are described in the following subsections.

4.6.1 Frequency and Time Division Multiplexing

Frequency division multiplexing (FDM) modulates each message to a different carrier frequency. The modulated messages are transmitted through the same channel and a bank of filters separates the messages at the destination; see Figure 4.17. The frequency band of the system is divided into several narrowband channels, one for each user. Each narrowband channel is reserved for one user all the time. FDM has been used in analog carrier systems of the telephone network. The same principle is used also in analog cellular systems where each user occupies one FDM channel. In this case we'll call it *frequency division multiple access* (FDMA) because the frequency division method is now used to allow multiple users to access the network at the same time.

A more modern method of multiplexing is *time division multiplexing* (TDM), which puts different messages, for example PCM words of different

Communication facilities generally contain more capacity than is required by a single user. It is economically feasible to share communication media and equipment to allow multiple users to share the cost.

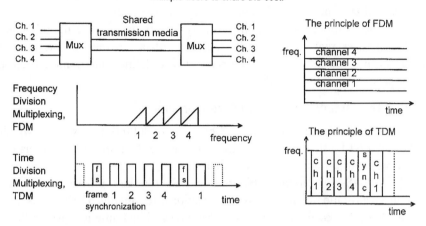

Figure 4.17 Multiplexing methods FDM and TDM.

users, in nonoverlapping timeslots. Each user channel uses a wide-frequency band but only a small fraction of time, which is called a timeslot; see Figure 4.17. In addition to the user channels, framing information is needed for the switching circuit at the receiver that separates the user channels (timeslots) in the demultiplexer. When the demultiplexer detects the frame synchronization word, it knows that this is the start of a new frame and the next timeslot contains the information of user channel 1.

This method of TDM is used in transmission systems such as optical line systems but also in digital cellular networks where we call it *time division multiple access* (TDMA). One user occupies one timeslot of a frame, and the time division principle allows multiple users to access the network at the same time using the same carrier frequency.

4.6.2 The PCM Frame Structure

We introduced the principle of TDM in the previous section. As an example of TDM we now look at the most common frame structure in a telecommunications network, namely, the primary rate 2048-kbit/s frame used in the European standard areas. This is the basic data stream that carries speech channels and ISDN-B channels through the network and it is called *E-1*. The corresponding North American primary rate is 1.544 Mbit/s, which carries 24 speech channels; it is known as DS1 or T1.

We explain here the 2048-kbit/s frame as an example because our objective is to get a clear view about TDM. The structure of the American T1 frame is explained later in this section and, for example, in [2].

In the European scheme the primary rate frame is built up in digital local exchanges that multiplex 30 speech or data channels at bit rate of 64 kbit/s into the 2048-kbit/s data rate. ITU-T defines this frame structure in the recommendation G.704.

4.6.2.1 The 2-Mbit/s Frame Structure (Europe)

PCM-coded speech is transmitted as 8-bit samples 8000 times a second, which makes up a 64-kbit/s data rate. These eight-bit words from different users are interleaved into a frame at a higher data rate.

The 2048-kbit/s frame in Figure 4.18 is used in the countries implementing European standards for telecommunications. It contains 32 timeslots, and 30 of them are used for speech or 64-kbit/s data. The frame is repeated 8000 times a second, which corresponds to the PCM sampling rate. Each timeslot contains an eight-bit sample value and the data rate of each channel is 64 kbit/s. These voice channels or data channels are synchronously multiplexed into a 2-Mbit/s data stream. For error-free operation the tributaries (64-kbit/s data streams of the users) have to be synchronized with the clock signal of the 2-Mbit/s multiplexer. The data rate of 2048 kbit/s of the multiplexer is allowed to vary ±50 ppm (parts per million), and as a consequence each user of the network has to take timing from the multiplexer in the network and generate data exactly at the data rate of the multiplexer divided by 32.

Frame Synchronization Timeslot

The frame alignment word is needed to inform the demultiplexer where the words of the channels are located in the received 2-Mbit/s data stream. A *frame*

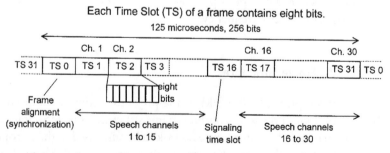

Each Time Slot (TS) of a frame contains eight bits.

One time slot (TS0) is used for frame synchronization and one (TS 16) for signaling. Data rate of frame: 8000 1/sec * 8 bits * 32 = 2048kbit/s

Figure 4.18 The 2048-kbit/s frame structure, G.704.

synchronization timeslot (TS0) includes frame alignment information and it has two different contents that are alternated in subsequent frames; see Figure 4.19. The demultiplexer looks for this timeslot in the received data stream and, when it is found, locks to it and starts picking up bytes from the timeslots for each receiving user. A fixed alignment word is not reliable enough for frame synchronization because it may happen that a user data of one channel simulates the synchronization word and the demultiplexer might lock to this user timeslot instead of TS0. This is why there is one alternating bit (D2) in timeslot 0 (see Figure 4.19) and, therefore, the demultiplexer is able to detect the situation where one channel transmits a word that is equal to the frame synchronization word.

To make frame alignment even more reliable, the *cyclic redundancy check 4* (CRC-4) procedure was added in the middle of the 1980s. C-bits are allocated to carry a four-bit error check code that is calculated over all the bits of a few frames. The receiver performs an error check calculation over all the bits of the frames and is able to detect false frame alignment even if the frame alignment word is simulated by one user that alters bit 2.

Each receiver of the 2048-kbit/s data stream detects errors in order to monitor the quality of the received signal. Error monitoring is mainly based on detecting errors in the frame alignment word. The receiver compares the received frame alignment word in every other TS0 with the error-free frame word. In addition to the frame alignment word, the CRC-4 code is used to detect low error rates in addition to frame alignment. Errors in the frame alignment word do not give reliable results when the error rate is very low. It may take a long time before an error is detected in TS0 even when errors have occurred in other timeslots of the frame.

The TS0 in every other frame contains also a far end alarm information bit A as shown in Figure 4.19. This is used to tell to the transmitting multiplexer that there is a severe problem in the transmission connection and reception is not successful at the other end of the system. This can be caused for example by high error rate, loss of frame alignment or loss of signal. With the help of the far end alarm consequent actions can take place. These may be, for example, rerouting user channels to another operational system.

The Structure of the Signaling Timeslot

Timeslot 16 (TS16) is defined for use with CAS to carry separate signaling information of all the user channels of the frame. For CAS the signaling information of all the users is carried in common data packets and any timeslot can be used. In this case the TS16 does not contain a multiframe structure.

TS16 is a transparent 64-kbit/s data channel like any other timeslot in the frame. Note that byte (word) synchronization is available in the interface

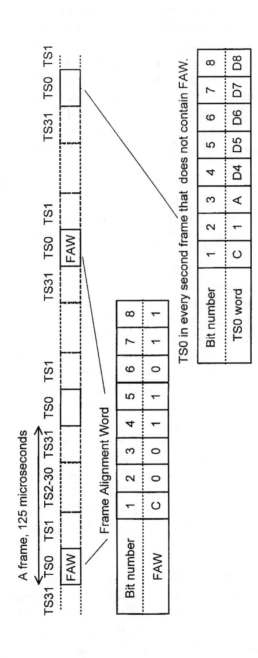

Figure 4.19 The 2048-kbit/s frame alignment word in TS0. A, far-end alarm, alarm condition "1"; D, spare bits that can be used for specific point-to-point low data rate applications; bit 2 alternates from frame to frame to prevent accidental simulations of the frame alignment signal; C, CRC-4 procedure for protection against simulation or frame alignment and enhanced error monitoring.

of a 2048-kbit/s multiplexer. Byte synchronization means that each timeslot contains the complete value of a sample, not for example the last four bits of the previous one and the first four bits of the next sample. The multiframe structure explained here relies on the precise knowledge of where each eight-bit byte begins.

There are 30 channels that share the signaling capacity of TS16. A frame structure is needed to allocate the bits of this timeslot to each of the 30 speech channels. The information about the location of the signaling data of each speech channel is given to the demultiplexer with the help of multiframe structure; see Figure 4.20.

The multiframe synchronization word 0000 is repeated in 2-ms periods and tells that the next eight-bit TS16 of the following frame contains signaling information for user channels 1 and 16, four bits for each of them. Thus the signaling data rate for each channel is 2 kbit/s. To avoid false multiframe alignment, the signaling sequence 0000 cannot be used.

The multiframe alarm bit, "A", is used to indicate far-end malfunction. If the demultiplexer is unable to detect multiframe synchronization, it sets the alarm bit to "1" in the opposite transmission direction in order to inform the transmitting end.

Note that the multiframe structure is completely independent of the frame structure in Figure 4.19 and the multiframe may start in TS16 of any 125-μs frame.

4.6.2.2 The 1.544-Mbit/s Frame Structure (United States)

The primary data rate in the United States and Japan is 1.544 Mbit/s instead of 2.048 Mbit/s used in the areas using European standards. As European PCM frame also 1.544-Mbit/s frame is repeated at PCM sampling rate that is 8000 times in a second. The frame structure shown in Figure 4.21 is used in North America and is known as the T1 or DS1 frame [2].

The North American PCM system accomplishes frame alignment differently than the European 2-Mbit/s system. Like its European counterpart, it uses eight-bit timeslots but each frame contains 24 channels. To each frame one bit is added, called a framing bit, and we get a 1.544-Mbit/s data rate, as shown in Figure 4.21.

As we see a timeslot is not reserved for CAS information as we had in the 2-Mbit/s frame structure. Instead, the least significant bit of each channel in every sixth frame is used for signaling. As a consequence, only seven bits in each timeslot are transparently carried through the network and basic user data rate is 56 kbit/s instead of 64 kbit/s of the European systems.

For frame synchronization and demultiplexing of signaling information frames make up a multiframe structure with two alternative lengths: a superfame

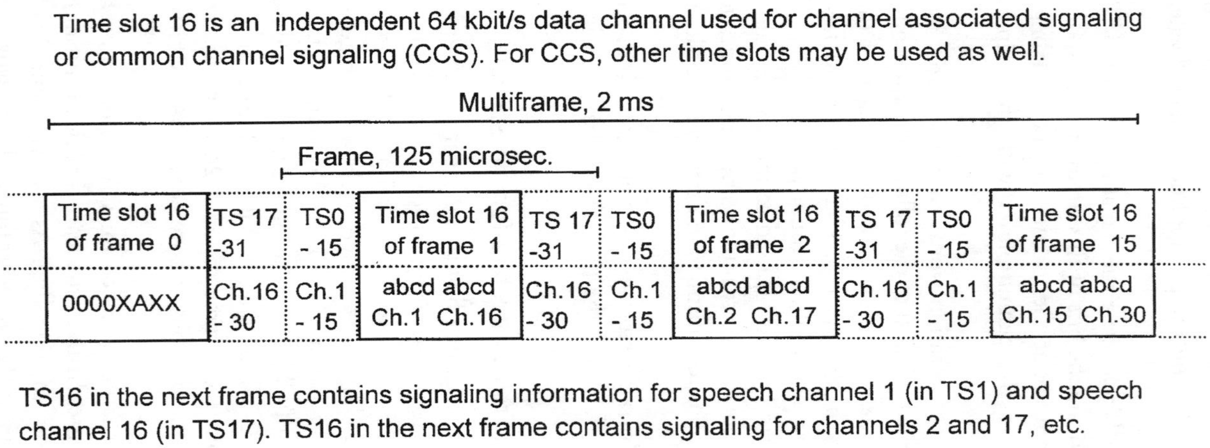

Time slot 16 is an independent 64 kbit/s data channel used for channel associated signaling or common channel signaling (CCS). For CCS, other time slots may be used as well.

Multiframe, 2 ms

Frame, 125 microsec.

Time slot 16 of frame 0	TS 17 -31	TS0 -15	Time slot 16 of frame 1	TS 17 -31	TS0 -15	Time slot 16 of frame 2	TS 17 -31	TS0 -15	Time slot 16 of frame 15
0000XAXX	Ch.16 -30	Ch.1 -15	abcd abcd Ch.1 Ch.16	Ch.16 -30	Ch.1 -15	abcd abcd Ch.2 Ch.17	Ch.16 -30	Ch.1 -15	abcd abcd Ch.15 Ch.30

TS16 in the next frame contains signaling information for speech channel 1 (in TS1) and speech channel 16 (in TS17). TS16 in the next frame contains signaling for channels 2 and 17, etc.

Figure 4.20 The multiframe structure of TS16. For CAS TS16 of 16 frames form a multiframe with four bit signaling information for each of the 30 speech channels. The multiframe contains synchronization "0000", remote alarm information "A" (alarm = 1) and spare bits "X" that are set to 1.

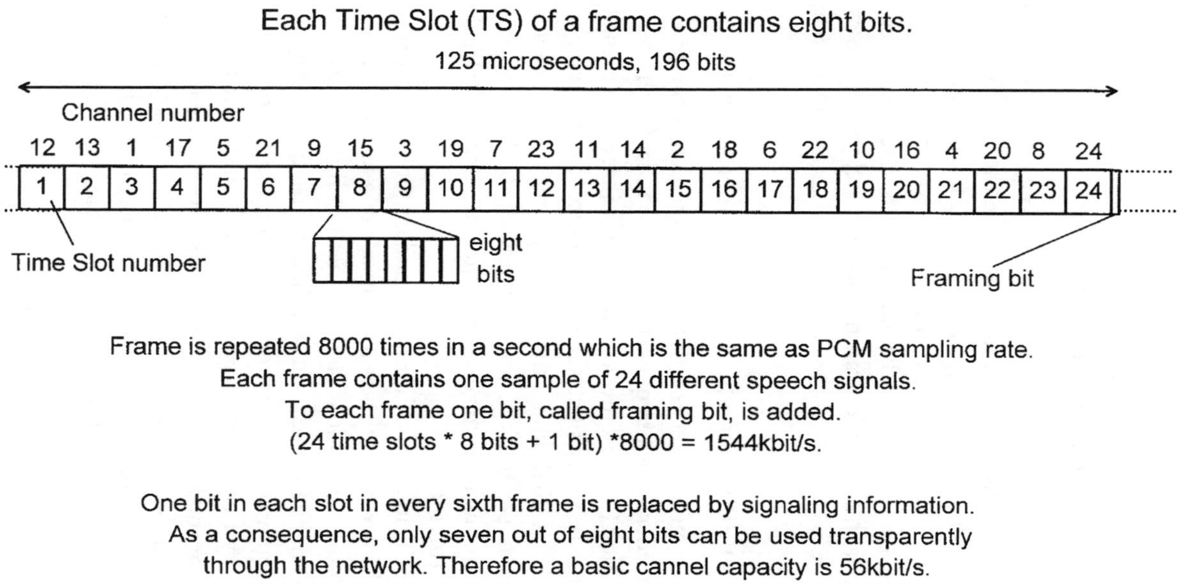

Each Time Slot (TS) of a frame contains eight bits.

125 microseconds, 196 bits

Channel number

12 13 1 17 5 21 9 15 3 19 7 23 11 14 2 18 6 22 10 16 4 20 8 24

1 2 3 4 5 6 7 8 9 10 11 12 13 14 15 16 17 18 19 20 21 22 23 24

Time Slot number

eight bits

Framing bit

Frame is repeated 8000 times in a second which is the same as PCM sampling rate.
Each frame contains one sample of 24 different speech signals.
To each frame one bit, called framing bit, is added.
(24 time slots * 8 bits + 1 bit) *8000 = 1544kbit/s.

One bit in each slot in every sixth frame is replaced by signaling information.
As a consequence, only seven out of eight bits can be used transparently
through the network. Therefore a basic cannel capacity is 56kbit/s.

Figure 4.21 The 1.544-Mbit/s PCM frame.

containing 12 frames or an *extended superframe* (ESF) continuing 24 frames. The framing bits, one in each frame, carry frame synchronization information including CRC-code. The detailed structure of the multiframe is explained, for example, in [2].

4.6.3 Plesiochronous Transmission Hierarchy

A primary rate of 1.5 or 2 Mbit/s is often too slow for transmission in trunk or even in local networks. This was noticed in the early 1970s and the ITU-T standardized the higher data-rate systems for transmission in the latter half of the seventies. The digital systems of those days were mostly carrying analog information and end-to-end synchronization was rarely required. The first standardized digital higher order transmission hierarchy is known as *plesiochronous digital hierarchy* (PDH). We review first the European hierarchy of higher order multiplexing.

4.6.3.1 European PDH Hierarchy for Higher Order Multiplexing

The higher order multiplexers of PDH are allowed to operate according to their own independent clock frequencies. These standards were based on plesiochronous operation ("almost the same data rate") that allows a small-frequency difference between tributary signals that are multiplexed into a higher aggregate rate. For example, at 2048 kbit/s the frequency tolerance was standardized to be ±50 ppm (parts per million) and at 8448 kbit/s the allowed tolerance is ±20 ppm. This means that, for example, the data rate of 2048-kbit/s system may deviate by 100 bit/s.

The basic principle of the European standard for higher order multiplexers is that each multiplexer stage takes four signals of a lower data rate and packs them together into a signal at a data rate that is a little bit over four times as high, as shown in Figure 4.21. In addition to tributaries, aggregate frames contain frame alignment information and justification information.

The tributary frequencies may differ slightly and their frequencies must be justified to the higher order frame. This process, called justification or stuffing, adds a number of justification bits to each tributary in order to make the average tributary data rates exactly the same. In the demultiplexer these justification bits are extracted and the original data rate for each tributary is generated.

At each hierarchy level the tributary signals are *bit interleaved* to the aggregate data stream. Additional bits are needed in the frame for frame synchronization (frame alignment) and justification, and therefore the next level has a slightly higher rate than four times the tributary rate. The frame also contains some spare bits that can be used, for example, for management

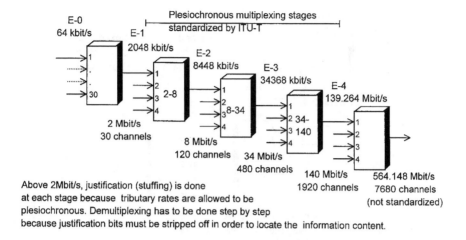

Above 2Mbit/s, justification (stuffing) is done at each stage because tributary rates are allowed to be plesiochronous. Demultiplexing has to be done step by step because justification bits must be stripped off in order to locate the information content.

Figure 4.22 The plesiochronous digital hierarchy (European standard).

data transmission for the Network Management System in use. Bits for far-end alarms are included in the frames, just as we had in the 2048-kbit/s frame discussed previously.

The standards for PDH ensure compatibility in multiplexing between systems from different manufacturers. The management functions are not standardized and differ from manufacturer to manufacturer.

Only the local interfaces and the multiplexing scheme are standardized in PDH. The multiplexers are connected for transmission via standard interfaces at 2, 8, 34, or 140 Mbit/s to separate line terminal equipment or to a higher order multiplexer, as shown in Figure 4.22. The line interfaces of the terminals for copper cable, optical fiber, and radio transmission are manufacturer-specific and the vendor has to be the same at both ends.

4.6.3.2 The North American PDH Hierarchy for Higher Order Multiplexing

The North American PDH hierarchy is shown in Figure 4.23. Higher order rates are DS1C (3.152 Mbit/s), DS2 (6.132 Mbit/s), DS3 (44.736 Mbit/s), and DS4 (274.176 Mbit/s) [2]. The higher-level multiplexers are named in such a way that we know the DS levels that are being combined. For example, M13 in Figure 4.23 has inputs from level DS1 and the output is at level DS3.

As we see in Figure 4.23 a higher order bit rate of each multiplexer is a little bit higher than the sum of the tributary data rates. The aggregate data stream at each level contains, in addition to tributary signals, framing information and the stuffing bits that are used to justify tributary data rates, which may have slightly different data rates, into the higher order frame. In the

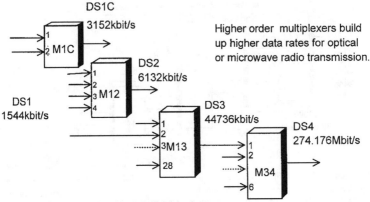

Higher order multiplexers build
up higher data rates for optical
or microwave radio transmission.

Above 1.5Mbit/s, justification (stuffing) is done
at each stage because tributary rates are allowed to be
plesiochronous. Demultiplexing has to be done step by step
because justification bits must be stripped off in order to locate the information content.

Figure 4.23 North American PDH digital hierarchy.

demultiplexer these stuffing bits are stripped off and the original tributary rate
is produced.

4.6.4 Synchronous Digital Hierarchy and SONET

The PDH higher order systems were standardized twenty years ago. By the
end of the 1980s a lot of optical fiber cable was installed and analog networks
were upgraded into digital networks. It was realized that new standards were
required to meet future requirements.

Problems with the PDH standards are:

- Access to a tributary rate requires step by step demultiplexing because
  of stuffing (justification).
- Optical interfaces are not standardized but vendor-specific.
- To use optical cables, a separate multiplexer for each level (e.g., multi-
  plexing from 2 to 140 Mbit/s requires 21 multiplexing equipment)
  and separate line terminals are needed.
- American and European standards are not compatible.
- Network management features and interfaces are vendor-dependent.

ANSI started to study a new transmission method in the middle of the
1970s to utilize optical networks and modern digital technology more effi-

ciently. This system is called synchronous optical network (SONET), and it is used in the United States.

ITU-T made its own worldwide standard, called *synchronous digital hierarchy* (SDH) by the end of the 1980s. SDH is actually an international extension to SONET and is based on SONET but adapted to European networks as well. A subset of SDH recommendations of ITU-T was selected as a standard for the European SDH by ETSI. We may say that there are two different synchronous optical systems: SONET used in the United States and SDH used in areas where European standards are adapted. Operation principles of SONET and European SDH are quite similar and they use the same data rate at some levels that are shown in Table 4.3.

Figure 4.24 shows data rates of European SDH as well as an example of SDH equipment. SDH is a standardized multiplexing system for both

Table 4.3
Data Rates of SONET (United States) and Corresponding SDH Data Streams (Europe)

OC-N Optical Carrier Level	STS-N Electrical Level	Data Rate Mbit/s	SDH STM-N
OC-1	STS-1	51.84	
OC-3	STS-3	155.52	STM-1
OC-12	STS-12	622.08	STM-4
OC-24	STS-24	1244.16	
OC-48	STS-48	2488.32	STM-16
OC-198	STS-192	9953.28	STM-64

SDH systems are combined multiplex and optical line systems with standardized high data rate optical interface, network management interface and add/drop, terminal and cross-connect functions.

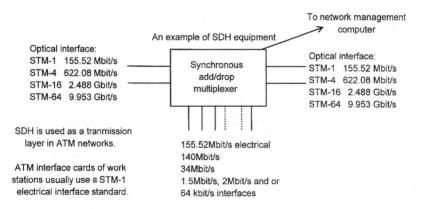

Figure 4.24 Synchronous digital hierarchy of ETSI.

plesiochronous tributaries (e.g., 1.5, 2, or 34 Mbit/s) and synchronous tributaries.

The main advantages that SDH provides over PDH standards are:

- The standardized high data rates for optical transmission (vendor-independence).
- Different systems are included in standards, for example terminal, add/drop, and crossconnection systems. These systems are discussed in Section 4.7 and make SDH-networks more flexible than PDH systems, which include only terminal multiplexer functionality.
- Efficient access to the tributary data rates.
- Tolerance against synchronization faults.
- Vendor-independent network management with sophisticated functions (in the future).

SDH aims to be the transmission standard for the future transport network. By transport network we mean the flexible high-capacity transmission network that is used to carry all types of information. By flexible we mean that telecommunication operators are able to easily modify the structure of the transport network from a centralized management system. This makes the delivery times of leased lines shorter. Leased lines are needed, for example, for LAN interconnections between the offices of a corporation. SDH is defined to provide broadband data transmission for ATM networks, which we will discuss in Chapter 6.

4.6.4.1 Multiplexing Scheme in SDH

The transmission data streams of SDH are called *synchronous transport modules* (STM), and they are exact multiples of STM-1 at a 155.52-Mbit/s data rate, as we see in Figure 4.22. STM-1 data is simply byte interleaved with other STM-1 data streams to make up a higher transmission data rate, and no additional framing information is added. Byte interleaving means that, for example, an STM-4 signal contains a byte (eight bits) from the first STM-1 tributary, then from the second, third, and fourth tributary, and then again from the first one. The demultiplexer receives all STM-1 frames independently.

The STM-1 frame is repeated 8000 times a second, which is equal to the PCM sampling rate. This makes each eight-bit speech sample visible in a 155.52-Mbit/s data stream. When PCM coding is synchronized to the same source as SDH systems, we can demultiplex one speech channel just by picking up one byte from each frame. The SDH frame contains frame alignment information and other information such as management data channels and pointers that tell the location of tributaries in the frame.

If tributaries are not synchronous with the STM-1 frame, a pointer (a binary number) in a fixed location in the STM-1 frame tells the location of each tributary. By looking at the value of this pointer we may easily find the desired tributary signal. This is a great advantage over PDH systems that require step-by-step demultiplexing (to separate information and stuffing bits) to the level of the tributary that we want to remove from the high-rate data stream.

Multiplexing in SDH is quite a complicated matter because the multiplexing supports very many different PDH and SDH streams to be multiplexed into an STM-1 stream. A more detailed treatment of the subject is not included here.

4.6.4.2 Data Rates of North American SONET

The *synchronous transport signal level 1* (STS-1) is the basic SONET module that corresponds to STM-1 of SDH. These modules have a bit rate of 51.840 Mbps and are multiplexed synchronously into higher order signals STS-N. Each STS-N signal has a corresponding optical signal called *optical carrier* (OC-N) for optical transmission. Table 4.3 presents data rates of SONET and the corresponding signal levels of European SDH.

An STS-1 signal consists of frames, and the frame duration is 125 μs (8000 times a second, which is equal to the PCM sampling rate) just as in SDH. Each frame contains 810 bytes, which makes up a bit rate of 51.840 Mbit/s. Transport overhead information such as frame synchronization and pointers use 27 bytes in each frame and the rest of it is used for payload, for example, 1.544-Mbit/s signals that contain PCM speech channels. The detailed multiplexing scheme of either SONET or SDH is not presented here; for more detailed information the reader is referred to [2].

4.7 Transmission Media

Transmission systems may use copper cable, optical cable, or a radio channel to interconnect far-end and near-end equipment. These channels and their characteristics are introduced in the following subsections.

4.7.1 Copper Cables

Copper cable is the oldest and most common transmission media. Its main disadvantages are high attenuation and sensibility to electrical interference. Attenuation in the copper cable increases with frequency approximately according to the formula

$$A_{dB} = k\sqrt{f} \text{ dB} \qquad (4.H)$$

where A_{dB} is attenuation in decibels, f is the frequency, and k is a constant specific for each cable.

This formula gives us the approximate attenuation at other frequencies if the attenuation at one frequency is known. For example, if we measure that attenuation of a certain cable is 6 dB at 250 kHz, then at a four times higher frequency of 1 MHz it is approximately 12 dB. The speed of propagation in a copper cable is approximately 200,000 km/s. There are three main types of copper cables, as are shown in Figure 4.25.

4.7.1.1 Twisted Pair

A twisted pair consists of two insulated copper wires that are typically about 1-mm thick. These two wires are twisted together to reduce external electrical interference and interference from one pair to another in the same cable. The twisted pair is symmetrical and the difference in voltage (or, to be more accurate, electromagnetic wave) between these two wires contains the transmitted signal. The twisted pair is easy to install, requires little space, and does not cost a lot. Twisted pairs are used in the telecommunications networks in subscriber lines, 2-Mbit/s digital transmission with distances up to 2 km between repeaters, and short-haul data transmission up to 100 Mbit/s in LANs.

4.7.1.2 Open-Wire Lines

The oldest and simplest form of a two-wire line uses bare conductors suspended at pole tops. The wires must not touch each other, otherwise a short circuit

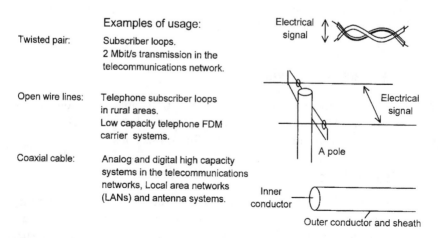

Figure 4.25 Transmission media, copper cables.

occurs in the line and communication will be interrupted. Open-wire lines are not installed any more, but they are still in use in many countries as subscriber lines or analog carrier systems with a small number of speech channels.

4.7.1.3 Coaxial Cable

A coaxial cable has stiff copper wire as the core that is surrounded by insulating material. The insulator is encased by a cylindrical conductor. The outer conductor is covered in protective plastic sheath. The construction of the coaxial cable gives a good combination of high bandwidth and excellent noise immunity. Coaxial cables are used in LANs, antenna systems for broadcast radio and TV, and in high-capacity analog and digital transmission systems in the telecommunications networks and even in older generation submarine systems.

4.7.2 Optical Fiber Cables

Optical fiber is the most modern media for transmission. It offers a wide bandwidth, low attenuation, and extremely high immunity to external electrical interference. The fiber optic links are used as the major media for long-distance transmission in all developed countries, and high-capacity coaxial cable systems are gradually being replaced by fiber systems.

An optical fiber has a central core (with a diameter from 5 μm to 60 μm) of very pure glass surrounded by an outer layer of less dense glass. A light ray is refracted from the surface between these materials back to the core and it propagates in the core from end to end. The principle of optical cable transmission and its main characteristics are presented in Figure 4.26.

The principle of optical fiber transmission has been known for some decades. The breakthrough of the technology was always expected to occur in the near future ever since the first half of the 1970s. However, the development of fiber manufacturing technology and optical component technology progressed more slowly than expected and the commercial breakthrough was delayed to the middle of the 1980s. Presently, all new high-capacity cable systems including submarine systems use optical fibers as a transmission medium.

The main advantages of optical fibers are:

- *High transmission capacity*: Optical fibers have a very large bandwidth and they are able to carry very high data rates, up to 50 Gbit/s.

- *Low cost*: The cost of the fiber has decreased to the level of a twisted cable pair; however, the coating and shielding of the cable increase the cost by a factor of two or more.

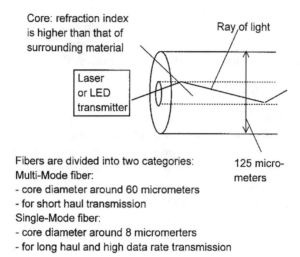

Core: refraction index
is higher than that of
surrounding material

Ray of light

Laser
or LED
transmitter

Fibers are divided into two categories: 125 micro-
Multi-Mode fiber: meters
- core diameter around 60 micrometers
- for short haul transmission
Single-Mode fiber:
- core diameter around 8 micromerters
- for long haul and high data rate transmission

Figure 4.26 Optical fiber.

- *Tolerance against external interference*: Electromagnetic disturbances have no influence on the light signal inside the fiber.

- *Small size and low weight*: Fiber material weighs little and the fiber diameter is only in the order of a hundred micrometers instead of a millimeter or more for copper cables.

- *Unlimited material source*: Quartz used in glass fibers is one of the most common materials on Earth.

- *Low attenuation*: Attenuation in modern fibers is less than a half decibel per kilometer and is independent of the data rate.

The disadvantages of optical fibers are that they are more difficult to install than copper cables and the radiation of light from a broken fiber may cause damage to the human eye. The safety standards restrict the allowable maximum optical power.

Fibers are divided into two main categories: multimode and single-mode fibers. Multimode fibers, with diameters of 125/62.5-μm cladding/core are used in short-haul applications like optical LANs, for example, *fiber-distributed digital interface* (FDDI). We will review LANs in Chapter 6. The typical attenuation of a multimode fiber is in the order of 2 dB/km.

Single-mode fibers, with diameters of 125/8 μm, are used in telecommunications network in high-rate and long-distance applications. The typical attenuation is in the order of 0.5 dB/km or less. (Compare the dimensions of optical fiber with the diameter of a human hair, which is approximately 100 μm.)

Optical transmission systems typically tolerate cable sections of tens of kilometers without intermediate repeaters. Long-haul, high-capacity coaxial cable systems usually require a repeater after every 1.5-km cable section! This partly explains the cost reduction of long-distance telecommunications since the 1970s.

4.7.3 Radio Waves

The most important advantage of radio transmission over cable transmission is that it does not require any physical medium. Radio systems are quick to install and, because cable does not require digging, the investment costs are low.

In telecommunications networks of today, microwave radio relay systems usually operate at radio frequencies from approximately 1 GHz to 40 GHz. These frequencies are focused with parabolic dish antennas and applicable communication distances range from a few kilometers up to approximately 50 km depending on the frequency in use and the characteristics of the system. The radio waves at these frequencies travel along a straight line, and therefore this kind of radio transmission is called line-of-sight transmission. The higher the frequency, the higher the attenuation, as we saw in Section 4.2, and the shorter the communication distance. At very high frequencies weather conditions influence attenuation and transmission quality, which restricts the available frequency band suitable for radio transmission. Figure 4.27 illustrates the structure of a point-to-point radio relay system used in the telecommunications network.

One important factor that restricts the usage of radio transmission is the shortage of frequency bands. The most suitable frequencies are already occupied.

4.7.4 Satellite Transmission

In satellite communications a microwave repeater is located in a satellite. An Earth station transmits to the satellite at one frequency band and the satellite regenerates and transmits the signal back at another frequency band. The frequencies used are in the range of 1 GHz to 30 GHz. Figure 4.28 illustrates the point to point transmission with the help of a geostationary satellite.

The satellites used in the telecommunications network are usually located in a so-called geostationary orbit so that they seem to be in the same location all the time from the point of view of the Earth station, as shown in Figure

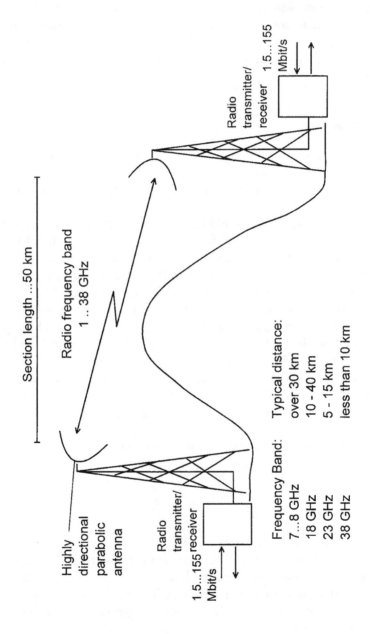

Section length ...50 km

Radio frequency band
1 .. 38 GHz

Highly directional parabolic antenna

Radio transmitter/receiver
1.5...155 Mbit/s

Radio transmitter/receiver 1.5...155 Mbit/s

Frequency Band:	Typical distance:
7...8 GHz	over 30 km
18 GHz	10 - 40 km
23 GHz	5 - 15 km
38 GHz	less than 10 km

Figure 4.27 Microwave radio transmission.

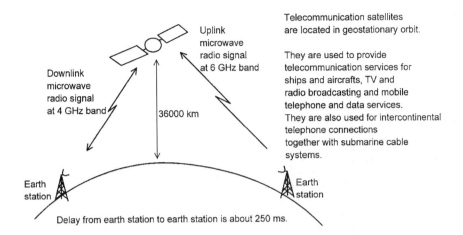

Figure 4.28 Satellite transmission.

4.28. The distance of this orbit is around 36,000 km from the Earth, which introduces a long transmission delay that is approximately 250 ms from the transmitting Earth station to the receiving Earth station. The speaker has to wait for a response for approximately 0.5 sec and this disturbs an interactive communication. Another problem is an echo that is also delayed by approximately 0.5 sec.

4.8 Transmission Equipment in the Network

There are many different systems needed in the telecommunications network to transmit signals via various different channels. We review the most common transmission devices or systems in this section. Some of them were already discussed in the previous section, and some of them are shown in Figure 4.29.

4.8.1 Modems

Modems are used to transmit digital signals over an analog transmission network or a transmission channel. Modems are described in Chapter 6 and are used to connect a *personal computer* (PC) or a terminal to an analog telephone channel. The microwave radio systems are also called modems because they send digital information over an analog microwave radio link, and in order to do this, they carry out modulation and demodulation processes.

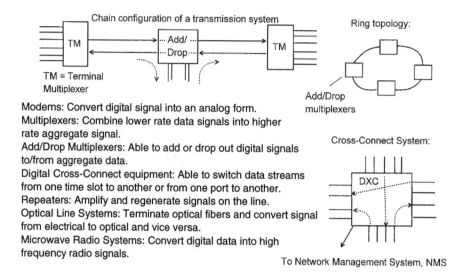

Modems: Convert digital signal into an analog form.
Multiplexers: Combine lower rate data signals into higher rate aggregate signal.
Add/Drop Multiplexers: Able to add or drop out digital signals to/from aggregate data.
Digital Cross-Connect equipment: Able to switch data streams from one time slot to another or from one port to another.
Repeaters: Amplify and regenerate signals on the line.
Optical Line Systems: Terminate optical fibers and convert signal from electrical to optical and vice versa.
Microwave Radio Systems: Convert digital data into high frequency radio signals.

Figure 4.29 Transmission equipment and system topology.

4.8.2 Multiplexers

Multiplexers combine digital signals to make up a higher bit rate for high capacity transmission. The digital multiplexing hierarchies in use are PDH, which is used in the present network, and SDH, which will gradually replace PDH systems in the future. These multiplexing schemes were described in Section 4.6.

4.8.3 Add/Drop Multiplexers

A transmission system in the network may be just a point-to-point system or it may be built as a chain or a ring system; see Figure 4.29. These configurations make efficient use of the system capacity feasible when only a small fraction of the capacity is needed on each equipment site. The add/drop multiplexers are used in these configurations to take out (drop) some channels from the high-rate data stream and add or insert other channels into it.

4.8.4 Digital Crossconnect Systems

The *crossconnect systems* (DXC) are network nodes that can rearrange channels in data streams. They make the network configuration of the transmission network flexible because, with the help of these nodes, a network operator is able to control actual transmission paths in the network remotely from the network management center. The basic functionality of DXC is the same as the functionality of digital exchanges that establish speech or ISDN connections. However, DXC is controlled by the network operator, not by a subscriber. There are crossconnect systems available that are able to switch high-order data rates, not only those of 56 or 64 kbit/s as ordinary exchanges do.

4.8.5 Regenerators or Intermediate Repeaters

Intermediate repeaters are needed if the communication distance is very long. They amplify an attenuated signal and regenerate the digital signal to the original form and transmit it further. The operation of a regenerator was described in Section 4.5.

4.8.6 Optical Line Systems

Optical line systems contain two terminal repeaters at each end of the fiber. They convert an electrical digital signal into an optical signal and vice versa. These systems include, as most other transmission systems do, supervisory functions like fault and performance monitoring. Note that SONET and SDH systems include multiplexing functions as well as the functions needed for optical transmission. In PDH multiplexers and optical line systems there are separate devices interconnected with standardized interfaces, which we discussed in Section 4.6.

4.8.7 Microwave Relay Systems

Microwave relay systems are radio systems that may be used for point-to-point transmission instead of copper or optical cable systems. They convert digital data into radio waves and vice versa. They also perform supervisory functions for remote performance and fault monitoring from the network management center.

4.9 High-Capacity Transmission Over Copper Cable Pairs

We have not yet utilized all the potential capacity of the most common transmission medium: symmetrical (twisted pair) copper cables. A new technology that is known as digital subscriber line (DSL) has become available. With the help of this technology we will be able to increase the data transmission rate over ordinary local loops to the order of several megabits per second in addition to the ordinary speech channel. This is far beyond the capacity of ISDN subscriber lines. The ISDN data channels are expensive dialed circuits that are switched by ISDN exchanges. In the case of DSL, data and speech are separated at the local exchange site. Then the data portion is connected to the data network, which provides more efficient and less expensive packet-switched data service, for example, for Internet access. We now review a few new DSL techniques and their applications.

4.9.1 Applications of DSL

The carriers or network operators are aiming their DSL services mainly at residential users. For them it provides high data-rate access to the Internet and at the same time ordinary telephone connection over a local loop.

The corporate network managers can also take an advantage of the benefits the new technology offers. For the interconnection of LANs between offices in the same region, the DSL offers a low cost and high data-rate option. It can in some applications replace more expensive frame relay, modem, or ISDN connections. Figure 4.30 illustrates three applications of DSL: remote access to data center, Internet access, and interconnection of LANs.

DSL replaces the ordinary local loop, and DSL modems are needed at both ends of the line. At the carrier's central office, the passband filter splits off the voice channel and routes it into the PSTN. A *DSL access multiplexer* (DSLAM) terminates the data channel and sends traffic onto an ATM, frame relay, or fixed data circuits, where it heads to a remote data center or the Internet.

The DSL is mainly designed to improve the utilization of subscriber cables in the access network. However, because it requires less intermediate

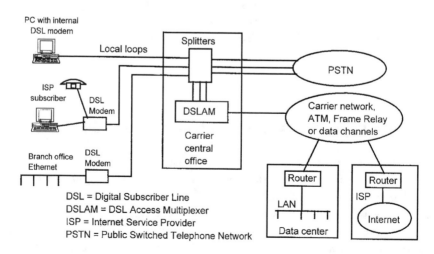

Figure 4.30 DSL in the local loop.

repeaters, system cost is reduced and the DSL is expected to some extent to replace conventional primary rate, 1.5- or 2-Mbit/s, copper cable transmission systems inside the network as well.

4.9.2 Digital Subscriber Line Techniques

The most important technologies and their data rates are presented in Table 4.4. We introduce these technologies here. Note that these technologies are evolving and the characteristics given in the table are not final. The data rates and distances in Table 4.4 are approximate figures based on present experience and are given for the comparison of the different DSL technologies. We may expect that some of the technologies introduced here will disappear and some of them will receive wide acceptance in a few years time.

The first digital subscriber line available for residential users was access to ISDN. It uses a single wire pair with 2 binary 1 quaternary (2B1Q) encoding along with TDM to create 2B + D basic ISDN access.

4.9.2.1 ISDN and Consumer DSL

For residential markets, some carriers in the United States plan to offer low-speed *ISDN DSL* (IDSL) and *consumer DSL* (CDSL). IDSL is ISDN without the ISDN switch. The two B channels of ISDN *basic rate interface* (BRI) are multiplexed to offer a dedicated 128 kbit/s of bandwidth for data only. This technology does not provide a simultaneous voice channel as other DSL technologies do. The CDSL, on the other hand, is a low-cost 1-Mbit/s version of DSL that provides a higher bit rate from the network to the subscriber.

Table 4.4
DSL Technologies and Service Rates

	Maximum rate, 6 km from the central office		Maximum rate 4 km from the central office	
	Downstream	Upstream	Downstream	Upstream
IDSL	128 kbit/s	128 kbit/s	128 kbit/s	128 kbit/s
CDSL	1 Mbit/s	128 kbit/s	1 Mbit/s	128 kbit/s
HDSL (two pairs)	1.544 Mbit/s	1.544 Mbit/s	1.544 Mbit/s or 2.048 kbit/s	1.544 Mbit/s or 2.048 Mbit/s
ADSL	1.5 Mbit/s	64 kbit/s	6 Mbit/s	640 kbit/s
SDSL	1 Mbit/s	1 Mbit/s	2 Mbit/s	2 Mbit/s
RADSL	1.5 Mbit/s	634 kbit/s	6 Mbit/s	640 kbit/s
VDSL	51 Mbit/s	2.3 Mbit/s	51 Mbit/s	2.3 Mbit/s

4.9.2.2 High Bit-Rate Digital Subscriber Line

A conventional primary rate transmission system "PCM system" operating at a 1.544- or 2.048-Mbit/s data rate over twisted pair copper cables uses two cable pairs, one for each transmission direction. In a typical cable, signal attenuation together with crosstalk (interference from other systems in the cable) restricts the transmission distance and a regenerator is required approximately after each 1.5-km cable section. These conventional 1.544- and 2048-Mbit/s systems use AMI and HDB-3 encoding.

The *high bit-rate digital subscriber line* (HDSL) increases section length and thus reduces the need for intermediate repeaters. This technology uses 2B1Q encoding that has superior spectral and distance characteristics. The HDSL uses two cable pairs and thus is not a consumer access technology.

HDSL systems use two cable pairs (ETSI has proposed also three-pair HDSL) for full-duplex transmission. The data rate is divided between pairs. In one pair, to one direction, it is only half of the data rate of conventional systems that use different cable pairs for each transmission direction. Further improvement is achieved with the help of an efficient line code. The line code in use is 2B1Q, which means that each pair of bits is coded into one quaternary symbol with four values to the line. This is the same line code that is used in ISDN basic rate subscriber lines for 160-kbit/s bidirectional transmission. Now one symbol with four values carries two bits of information. That reduces the data rate on the line to half of the binary rate, as we saw in Section 4.4. The lower transmission rate decreases attenuation and crosstalk. These developments together make transmission distance much longer than the distance of conventional systems. Table 4.5 presents the bit rates of the proposed system in one direction of each pair. The transmission distances without intermediate repeaters are also given in Table 4.5 for both two- and three-pair HDSL systems [3]. The data rates in the table contain overhead information and make up a 2.048-Mbit/s service rate that is divided between two or three pairs.

In the United States where first-order PCM systems have a transmission capacity of 1.5 Mbit/s, HDSL with two pairs and 784 kbit/s per pair is used. The HDSL system typically used two pairs and transmits the same data rate

Table 4.5
HDSL Data Rates and Transmission Distances

Number of Pairs	Bit Rate/Pair	Section Length
2	1168 kbit/s	4 km
3	784 kbit/s	4.6 km

to both directions just as conventional 1.5/2-Mbit/s copper cable transmission systems. It is expected to replace them to some extent because it requires less intermediate repeaters and reduces costs.

4.9.2.3 Asymmetrical Digital Subscriber Line

A symmetrical connection has the same data rate in both transmission directions. The conventional T1 and E1 (1.5- and 2-Mbit/s) transmission systems and HDSL systems are symmetrical in this sense. However, many applications do not require as much capacity from a subscriber to the network as from the network to a subscriber. One example of these applications is *video-on-demand* (VoD), which transmits one video program to a subscriber via an ordinary telephone subscriber pair. A subscriber needs only a narrowband channel to the network that enables him/her to select the video program. The *asymmetrical digital subscriber line* (ADSL) uses a single pair and transmits downstream at a high data rate and at a low data rate in the upstream direction. Figure 4.31 shows how the ADSL technique is used for VoD service.

ADSL was originally developed for VoD. This service has not (yet) been successful, but ADSL has ideal characteristics for the residential Internet users. An ordinary telephone call or an ISDN call and data (video in the case of VoD) transmission may take place at the same time. The ADSL terminals modulate the video signal and control signal to a higher frequency band that the telephone or ISDN basic rate signal does not use. The frequency band up to 410 kHz is in use and the transmission distance is restricted to approximately 5 km or 6 km in the case of a 2- or 1.5-Mbit/s data rate. Even higher data rates are supported by the ADSL technology and, for example, 6-Mbit/s transmission is possible up to a 4-km distance.

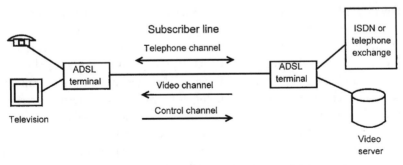

For a video data rate of 2 Mbit/s, the transmission distance may be up to 5 km.
The capacity of a control channel is 16 or 64 kbit/s.

Figure 4.31 Video-on demand and ADSL.

4.9.2.4 Symmetric DSL

The *symmetric digital subscriber line* (SDSL) system transmits at the same data rate to both directions just as HDSL but uses a single pair. Because both transmission directions operate at a high data rate, the near-end crosstalk is higher and the data rate lower than in ADSL; see Table 4.4.

4.9.2.5 Rate-Adaptive DSL

Rate-adaptive DSL (RADSL) is the latest DSL technology that is developed from ADSL. It is a very promising technology because it measures the performance characteristics of the local loop and dynamically adjusts to the highest speed possible. It can operate with symmetrical or asymmetrical send/receive channels. Asymmetrical configurations yield higher downstream data rates because of reduced near-end crosstalk (a low upstream data rate interferes less with downstream data). The RSDSL is easier to implement because it adapts itself to a wide range of loop conditions.

4.9.2.6 Very High Bit-Rate Digital Subscriber Line

The *very high bit-rate digital subscriber line* (VDSL) is a technology under development that aims to provide access to wider band services via ordinary telephone subscriber pairs than through other DSL technologies. The transmission range from the network to the subscriber's premises is up to 51 and 2.3 Mbit/s to the opposite direction. The distance over a cable pair without intermediate repeaters is quite short, in the order of a half kilometer. The distances in Table 4.4 require optical fibers in local loops that make it expensive to install. The copper-wire DSL part of the circuit might only include the drop line to a residence or business.

4.9.2.7 Summary of the DSL Technologies and Markets

Today most of the DSL products are proprietary, but standardization is proceeding in both ANSI and ETSI. The coding schemes of DSL systems as well as their higher level protocols differ from manufacturer to manufacturer. They may offer a direct Ethernet port for LAN applications and/or a serial port for other data and video applications. However, these technologies have important advantages over the competing technologies for high-speed Internet access such as the cable modems of the cable TV networks and the ISDN. There is a point-to-point local loop available that the DSL can utilize to provide access to a residence. It is straightforward to implement because there is a dedicated point-to-point line to each user. In the cable-TV networks we have to combine and split data from/to many users. The ISDN has a lower data rate and requires operator's investments to infrastructure to manage the increased load of the exchanges. The DSL removes traffic from the switched network and reduces the congestion that Internet users might cause. We may expect that DSL will succeed because it is flexible and compatible standards will evolve as products mature.

4.10 Problems and Review Questions

Problem 4.1: What is modulation and why is it used in transmission systems?

Problem 4.2: (a) Draw the spectrum of a cosine wave at the frequency of 1 kHz. (b) Draw the spectrum of an AM modulated signal when the carrier frequency is 100 kHz and the modulating message is a cosine wave at 1 kHz. (c) Draw the spectrum when the modulation method is SCDSB. (d) Draw the corresponding spectrum of SSB modulation.

Problem 4.3: (a) Draw the constellation diagram (or signal space diagram) for a PSK signal with eight possible carrier phases. Write in the diagram which bit combination each of them could represent. (b) Draw the constellation diagram of a 16-QAM signal. The signal (carrier) has sixteen combinations of phases and amplitudes. Select suitable bit combinations that these signal wave forms could represent.

Problem 4.4: Explain how the radio wave propagation mode differs at *low frequency* (LF), *medium frequency* (MF), and *ultra high frequency* (UHF) bands?

Problem 4.5: Estimate the transmission capacity of an optical fiber that operates over the 0.9- to 1.6-μm wavelength range if coherent optical transmission is used. Assume that the speed of light is the same as in space (300,000 km/s) and the following modulation methods are in use: (a) voice signal bandwidth is 4 kHz and it is SSB modulated into the fiber and (b) voice signal is PCM-coded and transmitted in a binary form through the cable. We assume that the modulation scheme in use is capable of transmitting 1 bit/Hz.

Problem 4.6: Derive the formula, $L = [(4\pi fl)/c)]^2$ for the free-space loss given in Section 4.2.6. Use the formulas for the capture area and a ball surface area given in Section 4.2.6.

Problem 4.7: Show that the equation of radio wave attenuation in decibels, $L_{dB} = 92.4 + 20 \log_{10} f/\text{kHz} + 20 \log_{10} l/\text{km}$ dB, follows from the equation of attenuation $L = [(4\pi fl)/c)]^2$.

Problem 4.8: The approximate distance between an Earth station and a geostationary satellite is 40,000 km. (a) What is the attenuation of the uplink radio section at 6-GHz frequency? (b) What is the attenuation in downlink direction at 4 GHz?

Problem 4.9: Consider a cell in a GSM cellular network operating at 900 MHz and a cell in a DCS1800 network operating at 1.8 GHz. The

DCS1800 base station is installed in the same site as the GSM base station. If we assume that all system parameters except the frequency are equal, what would be the radius of the DCS1800 cell if the radius of the GSM cell is 1 km?

Problem 4.10: How much higher transmission power is needed if the radio transmission distance is doubled (for the same performance)?

Problem 4.11: A telecommunications network operator is aiming to update a GSM network with DCS1800 base stations. The cells of GSM (900 MHz) are designed for the maximum transmission power of 1W. What should be the maximum transmission power of DCS1800 (1.8 GHz) base stations with the same cell structure?

Problem 4.12: What is the approximate gain of the satellite TV antenna when the diameter of the dish is 0.6m and the frequency is 10 GHz? How much better is the S/N if the diameter is 1m?

Problem 4.13: What are the theoretical maximum symbol rate r and binary bit rate C through the following baseband channels? (a) The bandwidth B = 3 kHz and the S/N = 20 dB (degraded speech channel); and (b) the bandwidth B = 5 MHz and the S/N = 48 dB (typical video channel).

Problem 4.14: How many bits can be coded into each symbol in the case of baseband systems (a) and (b) of the previous problem? (c) Explain how much higher the data rate is in the case of (b) because of wider bandwidth and how much higher the bit rate is because of the improved S/N compared with the channel in case (a) of Problem 4.12.

Problem 4.15: Estimate how many symbol values (crosses in the constellation diagram, see Figure 4.7) there should be in the case of a 28.8-kbit/s modem using QAM. The symbol rate of this modem is assumed to be 3 kBauds.

Problem 4.16: What is the purpose of coding in data transmission?

Problem 4.17: Explain how binary values "1" and "0" are represented in the following codes: (a) NRZ; (b) RZ; (c) AMI; and (d) Manchester.

Problem 4.18: Explain the operation principle of a regenerator (regenerative repeater).

Problem 4.19: What are the two main multiplexing methods and how do they operate?

Problem 4.20: Explain the structure of a 2-Mbit/s PCM frame.

Problem 4.21: Explain how CAS is transmitted in the 2-Mbit/s PCM frame.

Problem 4.22: Explain PDH.

Problem 4.23: What is SDH, and what advantages does it provide over PDH?

Problem 4.24: The measured attenuation at 1 MHz of a 1-km copper cable pair is 18 dB. What is the approximate attenuation at 250 kHz, 500 kHz, 2 MHz, and 4 MHz?

Problem 4.25: What are the advantages of (a) optical transmission, (b) microwave radio transmission, and (c) satellite transmission? Compare their characteristics.

Problem 4.26: Calculate the one-way and two-way delays of the transmitted signal from one Earth station to another Earth station via geostationary satellite. The distance between a satellite and each Earth station is assumed to be 40,000 km.

Problem 4.27: Explain the main applications and characteristics of the DSL systems.

References

[1]   Carlsson, A. B., Communication Systems, *An Introduction to Signals and Noise in Electrical Communication,* New York, NY: McGraw-Hill Book Company, 1986.

[2]   Freeman, L. F., *Telecommunication System Engineering,* New York, NY: John Wiley & Sons, 1996.

[3]   Ericsson Telecom, *Understanding Telecommunications,* Lund, Sweden: Ericsson Telecom, Telia and Studentlitteratur, 1997.

5

Mobile Communications

Radio telephones have been around for many decades but the capacity of these systems has been very limited. These networks consist of only a few BSs with which the mobile units communicate, and each *base station* (BS) covers a large geographical area. At one time the number of simultaneous calls inside the area covered by one BS was restricted to the number of channels available for this BS. Therefore, the capacity of these systems was low and the radio telephone service was available only for professionals. Some of these conventional mobile radio telephone networks are still in use.

During the 1970s, the development of digital switching and information technologies made modern cellular telephone systems feasible. The cellular principle offered a solution to the capacity problem. Different analog cellular standards were developed in Nordic Countries, the United States, and Japan at the end of 1970s.

In this chapter we introduce the idea and operation of cellular radio systems in general and then review some other mobile systems such as paging systems and cordless telephones. The common principles of cellular systems are valid for any *public land mobile network* (PLMN). In the last section of this chapter we review the structure and operation of GSM network. Our aim in this chapter is an understanding of what is required of the network to enable us to receive or initiate a call anywhere in the world. The natural requirement for this is that compatible service is available. In order to make this clear we use GSM as an example of digital cellular systems since it is presently the globally dominant digital technology.

5.1 Cellular Radio Principles

The main problem with conventional radio telephone networks was low capacity because of the limited frequency band available for this service. Cellular networks provide a solution for this by using the same frequencies in multiple areas inside the network. This principle of frequency reuse with the help of cellular network structure was invented in the Bell Laboratories during the 1960s. The technical development of radio frequency control, the microprocessor, and software technologies made cellular networks feasible by the end of 1970s.

The most important common characteristics of cellular systems are:

- Frequency reuse of cellular systems provides a much larger number of communication channels than the number of channels allocated to the system.

- Automated intercellular transfer, handover, ensures continuity of communication when there is a need to change BS.

- Continuous monitoring of communication between the mobile and BS to verify quality and identity the need for a cell transfer, handover.

- Automatic location of mobile stations within the network ensures that calls can be routed to mobiles.

- Mobile stations continuously listen to a common channel of the network in order to receive a call.

- The functions required of all telecommunication networks-such as operation, maintenance, and invoicing—are consistent with cellular systems.

Figure 5.1 presents the basic elements of a simplified cellular network. BSs are radio transmitter/receivers by which the *mobile stations* (MSs, such as telephones) are connected to the wire-line network. The BSs are connected to the *mobile switching center* (MSC) by primary rate digital connections. The MSC acts as a local exchange in the fixed network. In addition to the switching and other functions of an ordinary telephone exchange, the MSC also keeps track of the subscribers' locations with the help of location registers. We will discuss them in the following section.

Note that all cellular networks are designed to act as access networks. Their main purpose is to make mobile subscribers accessible from the global (fixed) telecommunications network. The mobile cellular networks always rely on a fixed network. They do not have a switching hierarchy similar to a fixed network (see Chapter 2). The calls from one cellular network to another and international calls are connected via a fixed PSTN.

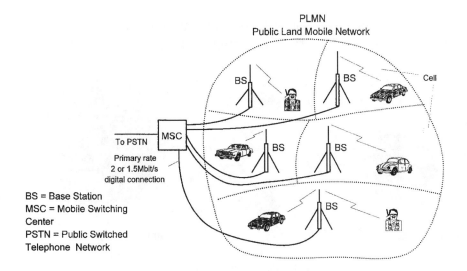

Figure 5.1 Structure of a cellular radio network.

5.2 Structure of a Cellular Network

This section will review the structure of a general cellular network. The detailed structure of a cellular radio network, the terminology of network elements, and their detailed functions are dependent on the network technology in question.

5.2.1 Cellular Structure

Instead of covering the whole area by high-power fixed radio stations the way older generation radio systems do, the area of a cellular network is divided into small *cells,* only a few kilometers or less across, as shown in Figure 5.2. As the subscriber density for the mobile telephones increases, the cell pattern can be changed to accommodate it. The power of the BSs and MSs is automatically decreased with the decreased cell size.

The BS and MS (telephone) are controlled to keep their transmission power as low as possible. This low power transmission does not interfere with other users of the same frequency (reuse of frequencies) some cells away from this cell. This is how each frequency channel may be used again and again and, in principle, a network operator may increase capacity without limit by reducing cell size; see Figure 5.2. This naturally requires investments for additional BS sites.

The consequences of reduced cell size are *handier* and *less expensive* telephones as well as *longer operational life* for the battery. Low transmission power

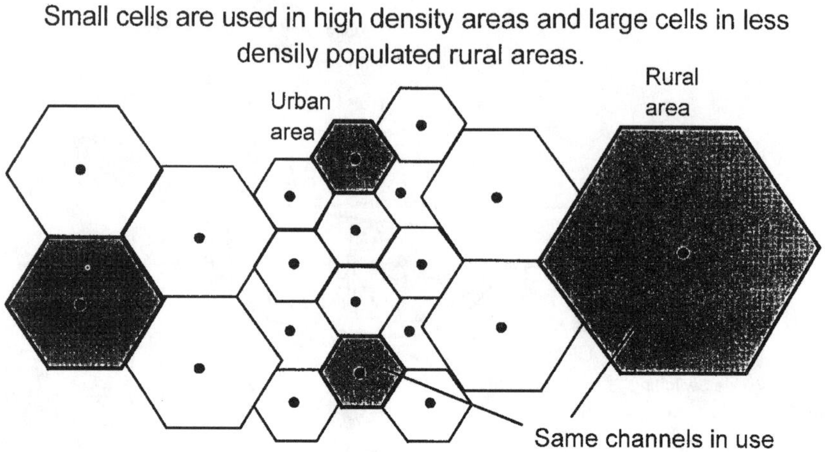

Small cells are used in high density areas and large cells in less densily populated rural areas.

Figure 5.2 Cellular structure of a mobile radio network.

also provides an improvement from the users' *safety* point of view. Because of public concern, this has become increasingly important now when most cellular telephones are handheld terminals. However, there is presently no evidence that radio waves cause harm to human health.

In a conventional fixed network, telephone calls are always routed to one fixed telephone socket, as we saw in Chapter 2. In a cellular mobile network a subscriber is located in one cell at a time. Now the network has to include additional intelligence to be able to connect a call to the cell where the called subscriber is available at that time. In order to perform this the cellular networks have two data bases, registers *home location register* (HLR) and *visitors location register* (VLR), and with them the network is able to manage the mobility of the subscribers.

5.2.2 Home and Visitors Location Registers

When a subscriber purchases a mobile telephone, she is registered in the HLR of her own mobile telephone operator. The HLR stores her up-to-date subscriber information such as where (in the area of which VLR) she is located presently, what services she has the right to use, and a number where she has transferred calls. The HLR is the global central point where her information is available wherever she is located. When a call must be routed to her, the dialed subscriber's telephone number tells the network where her HLR can be found.

VLR stores information of every subscriber in its area. The VLR informs the HLR when a new subscriber arrives in its area. It also contains more accurate information of where (to which cell or group of cells) to connect incoming calls directed to a certain subscriber. The VLR is usually integrated into a mobile telephone exchange, but the HLR is usually a physically separate efficient database system. The roles of the location registers are pointed out in Figure 5.3.

5.2.3 Radio Channels

Each BS provides two main types of channels, as shown in Figure 5.3. These are the common control channel and the dedicated channels. In downlink or forward direction, from network to mobile stations, information such as network identification, location information, designated power level, and paging (incoming call) is sent on the common control channel of each cell. When MSs are in idle mode (no ongoing call) they are continuously listening to the common control channel of one cell. In the uplink or reverse direction of the common control channel the MSs send, for example, call-request messages in the case of outgoing calls and location update messages when they notice that they have arrived in a new location area.

One dedicated user channel or a traffic channel is allocated for each call. During a call a MS is said to be in dedicated mode. Each dedicated channel requires the transmission of control information in addition to speech transmission. This is needed for transmission power control of mobile stations and for transmission of performance monitoring information from MSs to the network. When the call is cleared the dedicated channel is released and available for other users.

HLR, Home Location Register, stores subscriber information and updated
location information (VLR address). Each subscriber is registered in one fixed HLR.
VLR, Visitors Location Register, stores subscriber information of each
MS is located in its area.

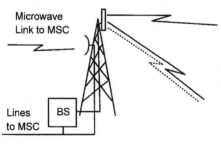

Microwave Link to MSC

Lines to MSC

BS

BS = Base Station
MSC = Mobile Switching Center

Common Control Channel that all mobiles listen to when they are in idle mode. Call request messages are sent on this channel in the case of an outgoing call and paging messages are sent in the case of an incoming call.

Dedicated channels that are used for calls (speech or traffic channels) when a mobile is in dedicated mode. Each dedicated channel has its own control channel for performance monitoring and control.

Figure 5.3 Base station and radio channels.

In Figure 5.3 we see that BSs are connected to the mobile switching center by a radio relay system or by a cable line (optical or copper cable). Especially in rural areas microwave links are attractive because cables are usually not available and an antenna tower is always needed for the BS antennas.

5.3 Operation Principle of a Cellular Network

In a fixed telephone network each subscriber is identified by a number of a subscriber loop that is connected to a certain telephone socket. In the case of a cellular telephone the identification is in the telephone set (MS) itself. The cell structure of the network and the mobility of the user require that the cellular network keep track of each MS in order to be able to route a call to its destination. We will now review the principles of how the cellular network manages the mobility of users and how calls are initiated and received. We will introduce the operation of a cellular network in general. The terms and operation presented may not be consistent with the terms and detailed operation of a certain network technology.

5.3.1 Mobile Station in Idle Mode

The MS is preprogrammed to know the frequencies of the control channels. When it is switched on, the mobile scans these frequencies and selects the BS with the strongest common control channel. Then the MS transmits its unique identification code, which may be its telephone number, over the control channel in order to inform the VLR. The VLR, with the help of the identification of the MS, finds out the address of the subscriber's home country and the home network. Then the MSC/VLR transmits the signaling message toward the home network. The message is then routed to the HLR, which is then informed that this specific subscriber is now located in the area of a certain VLR. The HLR stores this information so as to route the calls to the appropriate MSC/VLR, which then routes it further to the mobile subscriber.

The MS then continuously listens to the channel and, if necessary, transfers to the control channel of another cell; see Figure 5.4. Each network is divided into location or traffic areas that contain a group of cells. All BSs inside a certain location area send the same global code dedicated for that location area on the common control channel. If this location information sent by the network changes, the MS notices it and informs the network, which then updates the location information stored in the HLR and VLR.

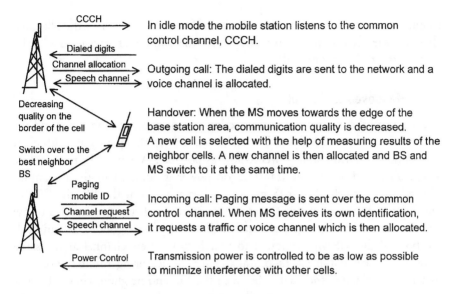

In idle mode the mobile station listens to the common control channel, CCCH.

Outgoing call: The dialed digits are sent to the network and a voice channel is allocated.

Handover: When the MS moves towards the edge of the base station area, communication quality is decreased. A new cell is selected with the help of measuring results of the neighbor cells. A new channel is then allocated and BS and MS switch to it at the same time.

Incoming call: Paging message is sent over the common control channel. When MS receives its own identification, it requests a traffic or voice channel which is then allocated.

Transmission power is controlled to be as low as possible to minimize interference with other cells.

Figure 5.4 The basic operation of the cellular network.

5.3.2 Outgoing Call

The number that a user wants to call is entered into the memory of the mobile telephone through its keypad. When the user presses the "call" button, the mobile telephone sends a signaling message to the BS via the common control channel. The message contains the dialed digits. The BS then passes them to the MSC for routing, as shown in Figure 5.4.

The MSC analyzes the dialed number, passes the digits to the public telephone network for call establishment through the PSTN, and requests a BS to allocate a dedicated speech channel for the calling mobile. The MS and BS switch to this channel. When the called party answers, the conversation may start, as shown in Figure 5.4.

5.3.3 Incoming Call

When a call is to be connected to the MS, the HLR informs to which VLR address the call should be routed. This address is global, containing the country and network codes. The call is then routed to the MSC/VLR, which knows the more exact location (the location area) of this specific subscriber inside its area. A paging message with MS identification is sent on the common control channel of all the BSs in that location area where the subscriber is presently located. The receiving MS continuously listens to this channel. When it receives the message containing its own identification, it requests a speech channel; a channel is then allocated for this call. The BS and MS switch to the allocated channel; the telephone alarms; and when the subscriber presses the "call" button, the call is connected.

5.3.4   Handover or Handoff

During a call the quality of the connection is continuously monitored and the transmission power of MS and BS is adjusted to keep the quality at a sufficient level and the transmission power as low as possible. When a MS moves close to the border of a cell, the transmission power is adjusted to the maximum allowed for that cell. As a MS moves further away from the BS, the S/N of the channel decreases and/or the error rate increases. If the quality falls below a predetermined level, a new channel is allocated in a neighbor cell and both the BS and the MS are requested to switch to the new channel at the same instant. The cellular network has analyzed the measuring results before the switch and estimated the quality between the MS and neighbor cells. The best alternative is selected for a new cell.

5.3.5   Mobile Station Transmitting Power

During the planning phase of a cellular network, the maximum transmitting power for each cell is defined. This power is dependent on the desired cell size and on geographical conditions. The transmitting power of the common control channel of the BS is adjusted to the level that is high enough to cover the cell area but not higher than necessary. During a call the transmitting power of a MS and a BS is continuously controlled by the network to minimize interference between cells that use the same frequency. This also saves the battery of the MS.

5.4   Mobile Communication Systems

So far we have looked at the generic operation of cellular mobile radio systems because of the importance of these systems. The cellular networks have been one of the fastest growing business areas in telecommunications during this 1990s. However, there are many other important mobile communication systems and we briefly introduce some of them in this section.

5.4.1   Cordless Telephones

Cordless phones were originally developed for the residential market and they were aimed to cover only one local area such as a house and garden. They support only local mobility and should not be understood as competitors for cellular mobile networks. We will now look at the most important applications of cordless telephones.

5.4.1.1 Residential Use

The only advantage of cordless telephones over fixed telephones in ordinary residential use is a wireless handset that allows some mobility. The BS of a cordless telephone is connected to the fixed telephone socket and only one handset for each BS is generally in use; see Figure 5.5. The BS unit contains a battery charger for the handset. Most systems in use are still analog first generation cordless phones, that are known as CT1, CT1+ systems.

5.4.1.2 Telepoint and Wireless Local Loop

The digital second generation cordless telephone technology (CT2) was developed for so-called telepoint use in addition to residential markets and offices. Telepoint is a service where BSs are installed in the key locations of a city such as railway stations, and airports. A user of this service may take his digital cordless telephone (or rent a cordless telephone) from home or the office and make a call outside via the telepoint BS. Subscribers are usually not able to receive a call. This service has not been successful, and most telecommunications network operators have laid it down due to the decreased cost of cellular mobile service, which allows better service and mobility.

With a new digital system called *Digital European Cordless Telecommunications* (DECT), the usage of cordless systems is developed toward a wider

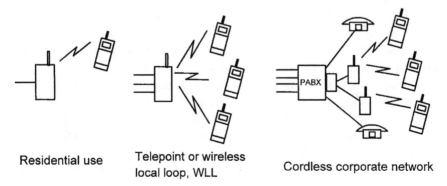

Applications have developed from residential use only to wireless local loop applications and wireless corporate networks with high rate data services. Modern digital systems (DECT, PACS and PHS) support all these applications.

Residential use   Telepoint or wireless local loop, WLL   Cordless corporate network

Figure 5.5 Cordless telephones and their applications.

coverage area such as the area close to the office and WLL applications. Both incoming and outgoing calls are supported.

The WLL applications are becoming increasingly important due to competition in the access networks that provide traditional fixed telephone subscriber services. With the help of this technology a new network operator may efficiently provide a service that is better, in terms of mobility, than the competing fixed telephone service by the operator who owns the cables of the fixed access network.

5.4.1.3   Cordless Corporate Network

In most companies internal wireless communications as well as external communications rely on the public cellular networks. The corporate telephone network is built on the fixed telephone service provided by the *Private (Automatic) Branch Exchange* (PBX/PABX) of a company. One attractive application of DECT may be cordless corporate networks where PABX is upgraded to control wireless DECT telephones in addition to wire-line telephones. This technology supports handover and terminals may move freely inside the area of one PABX that controls multiple BSs. Internetwork mobility management functions make it possible to extend the mobility of DECT to other office sites of a corporation and probably even to the local public network if the local public network operator supports DECT technology. Corresponding American technology is called a *personal access communications system* or PACS.

5.4.2   Private Mobile Radio

The *private mobile radio* (PMR) systems are dedicated and independent mobile radio systems. Some of them are just simple "walkie-talkie" type radios, others are complex networks that use a technology similar to public cellular mobile radio systems.

One typical PMR is owned by a taxi operator. It supports telephone calls (and probably some data communication) between a control desk and a number of car telephones in the area. A small number of radio channels are allocated for each of these systems inside each geographical area.

Traditionally each organization has built its own mobile radio system that is completely independent from others. The modern systems utilize a so-called trunking principle, which means that a group of radio channels is shared between several organizations. This improves the utilization of radio frequencies and is economically feasible because of reduced investments for network infrastructure. Each organization makes up a closed user group that operates the same way as if the systems were separate, but they are able to use any free radio channel.

The resource-sharing networks, also called trunked networks, are managed by a network operator to provide a specialized service for each virtual private network of a corporate customer. The use of frequency band is optimized by sharing it between multiple user organizations. The trunked networks will gradually replace conventional dispatch radio networks because of better spectrum efficiency and a wider range of services.

5.4.2.1 Principle of Operation of the Trunked Networks

In the trunked network, a central equipment allocates a free channel in real time to the user who requests it, and to him alone, for the duration of the communication. These modern systems utilize the same technology that is used in public cellular networks.

Because of this dynamic channel allocation the radio capacity of the network is utilized efficiently and users still feel as if they had a separate dedicated system in use. For each organization, closed user groups are defined in the system. Each user organization may have its own dispatch station just as in a separate conventional dispatch radio system.

Figure 5.6 explains the principle of channel allocation in the case of conventional PMR and in the case of trunked PMR. There are three conventional dispatch radio networks in this simplified example with one radio channel allocated for each, that is, each organization has only one channel available whether others communicate through their channels or not. There is a demand for three simultaneous calls, two for the organization 1 and one for the organization 3, and one of the calls is blocked although one radio channel is free.

The lower part in Figure 5.6 presents the principle of a trunked radio system. Now there is a pool of all three radio channels shared by all users.

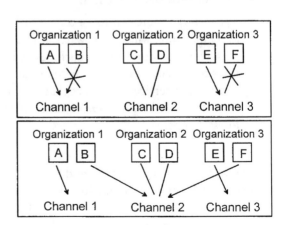

Figure 5.6 Operation principle of a resource sharing network, trunked network.

The channels in this pool are allocated on demand, and blocking occurs only when the total number of calls exceeds three in this simplified example.

5.4.2.2 Public Trunked Networks

There are many analog trunked networks in use. Examples of analog trunked networks are French Radiocom 2000 (providing a public cellular service as well) and Mobitex of Northern Europe and some other parts of the world. The frequency bands in use are different from those used in public cellular networks.

Most of the present analog trunked networks provide enhanced voice services such as:

- Group calling: A group is defined by the network operator, and each network typically contains a terminal that has a dispatching role;
- Priority calling in the case of an emergency;
- Call transfering;
- Conference calling;
- Voice messaging (with automatic retransmission in Mobitex).

The networks also support enhanced data transmission features such as:

- Predefined text messages (30 to 256 characters depending on the network);
- User-defined short messages;
- Telemetry, remote control of unmanned stations, measurements such as temperature, wind force, water level, and alarm collection of buildings; remote control: on/off machines, lights; status messaging: instructions to vehicles, respond of the status at the moment (taxi, bus, train, courier and service companies); and automatic vehicle location with help of the *Global Positioning System* (GPS), rescue, taxi, transport and forest companies.

The analog networks in use are different from country to country and there may even be many incompatible systems inside one country. There are new digital systems evolving which are aimed to support wider area service. One of them is the *Terrestial Trunked Radio* (TETRA) system, which is aimed to provide a compatible service in all European countries.

5.4.2.3 Terrestial Trunked Radio

A modern digital standard for European-wide PMR technology developed by ETSI is known as TETRA. This system was originally called Trans-European Trunked Radio and it is different from GSM but based on the experiences of it. It uses a different frequency band and provides some services that are not available in GSM, for example, mobile-to-mobile communication. The TETRA networks will be put into use by the end of the 1990s in Europe. These first networks will be built for public safety organizations such as police, fire brigades, and border guards. These systems use the 380- to 400-MHz frequency bands. Later the 410-, 450-, and 870-MHz frequency bands will be put into use by the commercial TETRA service for taxi, transport, railways, and other organizations.

As all trunked systems the TETRA uses a cellular network structure and channel allocation on demand to improve spectral efficiency. It is a modern digital system that further improves spectral efficiency.

Why do we need a separate network when the public cellular networks provide a service to define a closed user group for an organization? One reason is that the operation of the emergency services are so essential for a community that a separate network is required. The main reasons behind this are:

- Availability of capacity is independent of the activity of ordinary subscribers of the public cellular networks. In an emergency situation public cellular networks may become overloaded.
- Structure and services can be modified independently according to the users' needs.
- Some required features are not supported by the public cellular networks, for example, direct mobile-to-mobile communications and end-to-end encryption.

General features of TETRA systems are:

- Efficient use of spectrum, cellular structure, and trunked (shared) radio resources;
- Efficient use of investments, BSs, and exchanges shared between several organizations with national and international coverage (in the European Union);
- Standardized multivendor equipment;
- Support of *Virtual Private Networks* (VPN) for each user organization of the network; each of them may modify their resources, such as the usage of channels (mobile-to-mobile, mobile-to-BS) and priorities;

- Each user organization has its own "dispatcher station" from which an operator can communicate will all users;
- A number of channels may be permanently or temporarily allocated for a certain organization (quarantined share of recourses);
- Open channel (mobile-to-mobile and point-to-multipoint) communication supported;
- Prioritization of organizations and user groups;
- 28-kbit/s maximum data rate, packet- or circuit-switched.

The standardization of TETRA took place after the GSM and the requirements were slightly different than those for public cellular networks. The standardization work was carried out by the ETSI that had also specified the GSM. For these reasons, the technology of the TETRA is closely related to the GSM but differs in details.

Some key technical specifications of the TETRA are:

- TDMA (and FDMA) channel access method;
- *Frequency division duplex* (FDD);
- 25-kHz carrier spacing;
- Four-user channels per carrier;
- 28-kbit/s maximum user bit rate (all timeslots of one carrier used by a single user);
- Speech coding at the 4.8 kbit/s data rate.

5.4.3 Radio Paging

There are two basic types of radio paging networks: on-site pagers and *wide-area pagers* (WAP). On-site pagers cover a local area like a building or a hospital. WAP cover a wide area or even a whole country. Paging systems are used to transmit short texts or simply an audible beep. Pagers are small and inexpensive wireless communication devices that are used to receive notifications to the subscribers without disturbing their present activities.

The European Commission decided at the end of the 1980s to open up the paging system market and requested ETSI specify a new European coverage digital paging system that would support multiple operators. The 169-MHz frequency band was reserved for this service. In 1990 sixteen operators in eight European countries signed a formal agreement relating a timetable for setting up the ERMES system. The specifications were completed in 1992 and the service became available in the middle of the 1990s.

5.4.4 Analog Cellular Systems

In Section 5.3 we introduced the operation of cellular networks in general. The first cellular technologies were analog and they became available in the first half of the 1980s. These systems are often referred as the first generation cellular systems, and the most important analog cellular systems are:

- *Advanced Mobile Phone System* (AMPS) used in the United States;
- *Nordic Mobile Telephone* (NMT) used in Nordic Countries;
- *Total Access Communications System* (TACS) used in the United Kingdom.

These systems are quite similar but incompatible. They use a 800- to 900-MHz frequency band (NMT uses 450 MHz as well) and frequency modulation. The frequency band is divided into channels, and one of these is allocated for each call. We call this radio access principle FDMA.

5.4.5 Digital Cellular Systems

In this section we review the most important digital cellular networks that are in use today and some that are under standardization. We often refer to these systems as second generation cellular systems.

5.4.5.1 Global System for Mobile Communications

GSM operates at the 900-MHz frequency band and is presently the most widely used digital cellular technology. The structure and operation of the GSM network are explained in Section 5.6. In GSM the subscription information is stored on a smart card and a subscriber may change a mobile telephone any time. When he inserts his card to the new telephone he has access to exactly the same service as previously. The access method used in GSM is known as TDMA, where each frequency channel is divided into timeslots for multiple users.

5.4.5.2 Digital Cellular System at 1800 MHz

Digital cellular system at 1800 MHz (DCS1800) is also known as GSM1800 or the *Personal Communications Network* (PCN). It is based on GSM technology but operates at a 1710- to 1880-MHz frequency band and provides much higher capacity than GSM in terms of the number of users. DCS1800 is a technology for the European implementation of PCN, but it is in use in other parts of the world as well. The PCN is aiming to provide a mass mobile telecommunication service in urban areas. By the term "personal communications" we want to point out that each call is routed to a person instead of a certain fixed location, which is the case in the conventional fixed networks.

5.4.5.3 Personal Communications Network and Service

Personal communications refer to cellular mobile communications where a call is routed to a person who carries a terminal instead of a fixed terminal location as in the conventional fixed telephone network. The PCN and *Personal Communications Service* (PCS) mean simply microcellular systems that emphasize low-cost and high-capacity cellular service and a hand-portable terminal with a long battery life. In Europe the DCS-1800 is also called PCN because it is the implementation technology for PCN.

In the United States several digital technologies are used to implement the high-capacity cellular service that is known as PCS. These technologies are PCS-1900 (DCS-1800 at 1900 MHz), NADC (known also as D-AMPS or US-TDMA), and CDMA. All these network technologies are briefly introduced in the following subsections.

5.4.5.4 Personal Communications Service 1900

Personal Communication Service 1900 (PCS-1900) is based on GSM/DCS1800 technology but adapted to the frequency allocation of North America. These three GSM-based systems are so similar that with the help of a multimode MS a subscriber may use all these networks with the same terminal and subscription (same subscriber card). We will illustrate the structure and operation of this system in Section 5.6 as an example of a modern digital cellular system.

5.4.5.5 North American Digital Cellular

The United States and Canada have implemented digital techniques to increase the capacity and quality of the existing AMPS system. The *North American Digital Cellular* (NADC) system implements digital radio communication onto the frequency band of the AMPS. It divides channels of the analog AMPS into six timeslots. With the help of time division three or six (half-rate speech mode) users each share an analog 30-kHz AMPS channel. The terminals with dual-mode capability use a digital system when it is available, otherwise the analog AMPS service is used [1]. Because of this principle, the NADC system is also known as a *dual-mode AMPS* (D-AMPS).

The NADC network system is able to provide service for even the oldest analog AMPS terminal. The common control channels of analog AMPS and NADC are compatible and a MS, analog or dual-mode, first searches the *forward* (downlink) *control channel* (FOCC), which occupies one FDMA channel. Then the terminal informs the network with a signaling message that contains the information about its capabilities. There are presently three enhanced modes of operation in addition to the original analog AMPS:

- *Narrowband AMPS* (NAMPS): an analog enhancement designed by Motorola, which increases the capacity of the system;
- CDMA;
- NADC, which divides the analog 30-kHz channels into timeslots (TDMA).

Upon recognizing the mobile's enhanced digital capabilities the network will assign a *digital traffic channel* (DTC) to the mobile for a call. If a DTC is not available, then an analog channel is assigned instead. The DTC is defined by a channel number (frequency), timeslot number, a timing advance value, a cell-identifying code, and mobile power setting. The channel is maintained until the disconnect time with the help of the continuous quality measurement of the communication and handoffs or handovers if required.

5.4.5.6 Code Division Multiple Access

The CDMA has been selected by the most important network operators to become the main digital cellular standard in the United States. The main difference between CDMA and other technologies discussed previously is that on the radio path it does not use either FDMA or TDMA. Instead the mobiles use the wide frequency band all the time with the help of a unique code for each user. This unique code is used to spread the signal over a wide frequency band and to detect the wanted signal at the receiving end. The American system is also referred to as *narrowband CDMA* (N-CDMA) or *interim standard-95* (IS-95) system [1].

Operation Principle of the CDMA

The operation principle of CDMA radio transmission is not as easy to understand as FDMA or TDMA. Figure 5.7 shows a very simplified diagram of a CDMA system. In Figure 5.7 the spreading code data rate is ten times higher than the information data. In the actual IS-95 system the code has more than a hundred times higher data rate than the user data. The "exclusive-or" operation is then performed in the transmitter with the original data and the spreading code. The exclusive-or operation gives a high state when the data and code have different states and a low state when they are equal. In our simplified example the bits or symbols of the code and data+code are then ten times shorter than the bits of the original data. We saw in Chapter 4 that the shorter the pulses the wider spectrum they have. Thus the spectrum of each bit is now

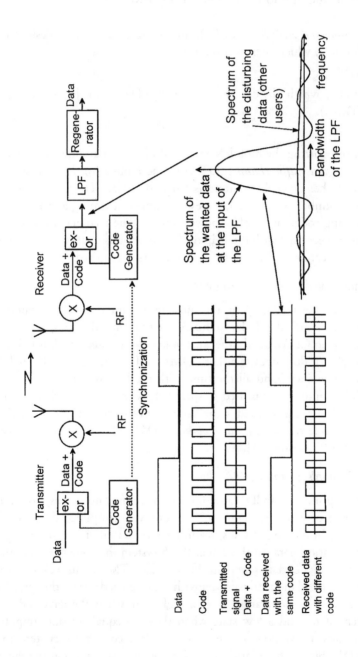

Figure 5.7 Operation principle of the CDMA.

ten times wider and the spectrum of the original data is spread over a ten times wider frequency band. After modulation with the RF-carrier a wide frequency band is occupied with this CDMA radio signal.

In the receiver the received signal is first demodulated and then the same code is used to detect the wanted signal at the other end. The same exclusive operation is performed in the receiver. The resulting data at the input of the low-pass filter in the receiver is the original low-rate data. We may imagine that the receiver, using the right code, has collected the signal energy from the wide frequency band to the baseband. The other signals (of other users) on the channel have been generated with different codes and are received as a random high data-rate signal with a wide spectrum, as shown in Figure 5.7. Most of these disturbing signals are filtered out in the low-pass filter (LPF) of the receiver, while most of the wanted low-rate data gets through the LPF. At the output of the LPF other signals are seen as noise on top of the wanted data. The regenerator detects the original data and this detection is error-free if the noise is not too high.

For proper operation the receiver has to be accurately synchronized with the transmitter and the simultaneously used codes have to be selected to minimize interference. The CDMA also requires accurate and frequent adjustment of the transmission power levels because the power of the users influence the S/N and the error rate of the other simultaneous users. The CDMA principle provides many advantages compared with FDMA or TDMA systems. It utilizes radio resources more efficiently and is not sensitive to multipath fading and narrowband radio disturbances. It uses low continuous transmission power, and the safety risk for the users of handheld phones is reduced.

IS-95 CDMA System

The CDMA system supports dual-mode operation just as another American system NADC. The CDMA resources exist in the same frequency band with the traditional AMPS system and occupies 41 AMPS channels (1.23 MHz). The CDMA users use their unique codes to share this frequency band. The common control channels are also spread over the CDMA band by their own spreading codes that are known by the MSs. When a MS is in the idle mode it uses the code of the FOCC (forward control channel) to listen to the network and to be able, for example, to receive a paging message in the case of an incoming call. When a call is connected, a new code is allotted to the user for the dedicated speech communication.

The CDMA is an interesting technology and it provides many other features that we have not discussed, such as soft handover or handoff. To perform soft handover a MS may use more than one BS (operating at the same CDMA frequency) at the same time with the different codes. When the need for handoff is detected, the connection is first established to the new BS and the old connection is released when the new channel is operating properly.

We restrict our discussion about CDMA and other cellular systems to brief introductions of the most important networks. For further information about the American CDMA the reader may refer to [2].

5.4.5.7 Japanese Digital Cellular

The *Japanese digital cellular* (JDC) system is also known as a *personal digital cellular* (PDC). It is a separate system from the previous analog but utilizes dual-mode terminals that are able to use an existing analog system as well. The network technology is very close to the European GSM.

5.4.5.8 Universal Mobile Telecommunication Service

As we have noticed, we have many different mobile networks in use today such as paging, cordless, analog cellular, and digital cellular systems. They are based on different standards and provide incompatible services. The *Universal Mobile Telecommunication Service* (UMTS) is a European concept for integrated mobile service in the future. It aims to provide a wide range of mobile services wherever the user is located. UMTS will use cordless, cellular, and satellite communications depending on the environment and the service required. UMTS is developed together with IMT2000. The cellular radio access method of UMTS will be wideband CDMA, which was approved by ETSI in the beginning of 1998.

5.4.5.9 International Mobile Communications 2000

The *International Mobile Communications 2000* (IMT2000) aims to be a global system for third generation mobile communications. It is being developed by the ITU, which was previously called *Future Public Land Mobile Telecommunications Service* (FPLMTS). It aims to provide the future global integrated mobile service. It will be quite similar to UMTS, but many problems will be incurred during the development of an integrated global system. Among them are frequency allocation and different political interests. There will be different systems in use in the future, but the development of mobile terminal technology will partly solve this problem for the users. The same terminal will be able use many different networks and the services that they provide.

5.4.6 Mobile Satellite Systems

One application of satellite communications was the point-to-point transmission presented in Section 4.6. Satellites presently provide mobile communications services to ships and aircrafts. They are used also in desert areas where other mobile services are not available. In the present systems that use geostationary satellites, MSs are expensive and the cost of service is quite high. There are many plans to implement lower cost satellite services that could be used in the future with handy MSs. These stations could probably use the same terminals that we will use in PLMN, such as GSM or CDMA. The two best known examples of these systems are Iridium of Motorola and Intelsat ICO.

These systems use many satellites that are placed in the orbit at a 700- to 10,000-km distance from the Earth instead of geostationary orbit with a 36,000-km distance. The satellites are circulating around the Earth in such a way that a few of them are visible all the time from any point on the Earth surface; see Figure 5.8. Each of the satellites performs BS functions and takes care of the large cell below it.

These systems need and use similar functions as cellular networks. Examples are the mobility management and handover necessary to manage the movement of satellites (BSs) instead of subscribers. These systems have many Earth stations that control the operation of satellites and behave as connection points to the public land networks.

In the future we will probably be able to use the satellite systems with multimode terminals that first search for the PLMN, for instance GSM. If it is not available a more expensive satellite service will be used.

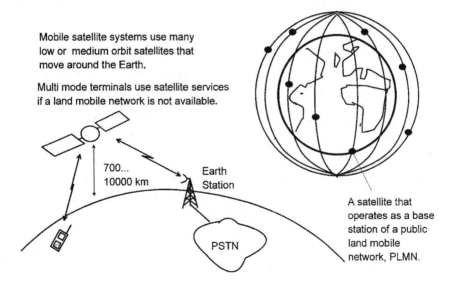

Mobile satellite systems use many low or medium orbit satellites that move around the Earth.

Multi mode terminals use satellite services if a land mobile network is not available.

700... 10000 km

Earth Station

PSTN

A satellite that operates as a base station of a public land mobile network, PLMN.

Figure 5.8 A mobile satellite system.

5.5 Global System for Mobile Communications and PCS-1900

As an example of a modern digital cellular network we introduce the structure and operation of a GSM network. The European digital cellular system GSM was developed by *Conference Europeen des Postes et Telecommunication* (CEPT) during the 1980s, and this work was continued by *European Telecommunication Standards Institute* (ETSI). The acronym GSM originally came from the standardization working team, but GSM is presently understood as GSM communications. There are two other cellular networks that are based on the GSM technology; the European DCS1800, which operates at a 1.8-GHz band, and the American PCS-1900, which operates at a 1.9-GHz band. Our discussion in this section is valid for all of these networks.

In GSM, unlike in analog mobile networks, subscription and mobile equipment are separated. Subscriber data is stored and handled by a *subscriber identity module* (SIM), which is a smart card belonging to a subscriber. With this card the subscriber may use any mobile telephone equipment just as if it was his own.

5.5.1 Structure of the GSM Network

A simplified diagram of GSM network architecture is presented in Figure 5.9. For a more detailed structure and functionality of the network the reader may refer to [1] or [2].

5.5.1.1 Radio Network

MSs are connected to the MSC via a *BS subsystem* (BSS). The BSS consists of a *BS controller* (BSC), and many *base transceiver stations* (BTSs) that are controlled by one BSC. The roles of the network elements are introduced in the following subsections.

5.5.1.2 Mobile Services Switching Center

As any local exchange, the MSC establishes calls by switching the incoming channels into outgoing channels. It also controls the communication, releases connections, and collects charging information.

As a mobile switching system, the MSC together with the VLR performs additional functions such as location registration and paging. It also transfers encryption parameters, participates in the handover procedure when required, and supports short message service. The short message service is a service integrated into GSM that enables users to transmit and receive short text messages.

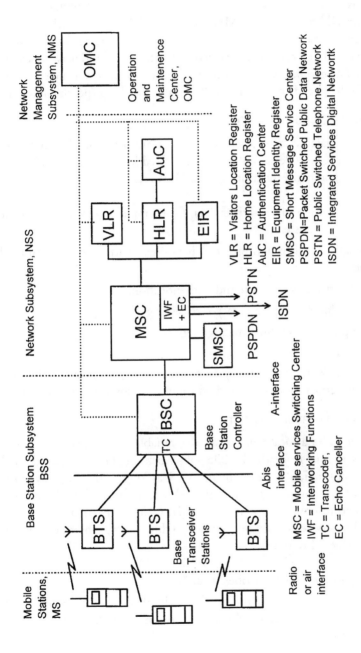

Figure 5.9 The structure of the GSM network.

In each cellular network there is at least one gateway MSC that provides connections to other networks. The MSC in Figure 5.9 performs gateway functions in addition to other MSC functions. The gateway MSC works as an interface between the cellular network and the fixed networks and must handle the signaling protocols between the fixed networks and network elements of PLMN. The gateway MSC also controls echo chancellers, which are needed between the fixed and cellular networks because of a long speech coding delay.

5.5.1.3 Home Location Register

All subscriber parameters of each mobile user are permanently stored in the HLR. The HLR provides a well-known and fixed location for variable routing information. The main functions of the HLR are:

- Organization of subscriber data, for example, service profile, services, and call transfer;
- Location registration and call handling, central store for subscriber location data;
- Support for encryption and authentication;
- Handling of supplementary services (e.g., barring);
- Support for the short message service.

The HLR is implemented by an efficient real-time database system that stores the subscriber data of a hundred thousand subscribers or even more.

5.5.1.4 Visitor Location Register

The VLR provides a local store of all the variables and functions needed to handle calls to and from the mobile subscribers in the area related to that VLR. The information is stored in the VLR as long as the MS stays in that area. The VLR communicates with the HLR to inform of the location of a subscriber and to obtain subscriber data that includes information about, for example, what services should be provided to this specific subscriber. The main functions of the VLR are:

- Organization of subscriber data;
- Management and allocation of the local identity codes in order to avoid frequent use of global identity on the radio path for security reasons;
- Location registration and call handling;
- Authentication;

- Support of encryption;
- Support for handover;
- Handling of supplementary services;
- Support short message service.

The VLR is a database system that is usually integrated in each mobile exchange, MSC.

5.5.1.5 Authentication Center

The security data of a subscriber is stored in the *authentication center* (AuC), which contains a subscriber-specific security key, encryption algorithms, and a random generator. The AuC produces security data with defined algorithms and gives subscriber-specific security keys to the HLR, which distributes them to the VLR. A PLMN may contain one or more AuCs, and they can be separate elements or integrated to the HLR. The same subscriber-specific key and algorithms are stored in SIM. There is no need to send them over the network and on the radio path.

5.5.1.6 Equipment Identity Register

The *equipment identity register* (EIR) is a database containing the information about mobile terminal equipment. There is a white list for serial numbers of the terminals that are allowed to use the service, a gray list for terminals that need to be held under surveillance, and a black list for stolen and faulty mobile terminals. Those terminals that are found on the black list are not allowed to use the network.

5.5.1.7 Interworking Functions

The *interworking functions* (IWF) is a functional entity associated with the gateway MSC. It enables interworking between a PLMN and a fixed network, an ISDN, a PSTN, and a PSPDN. It is needed, for example, in the case of high-rate data transmission, to convert a GSM internal digital transmission into a modem transmission inside the PSTN. It has no functionality with the service that is directly compatible with that of the fixed network.

5.5.1.8 Transcoder

A *transcoder* (TC) is needed to make a conversion between GSM voice coding (13 or 7 kbit/s) and PCM coding (64 kbit/s) that is used in the fixed network. In the case of data transmission, transcoding is disabled.

5.5.1.9 Echo Canceller

The *echo canceller* (EC) is needed at the interface between a GSM network and the PSTN. The efficient speech coding of GSM introduces such a long delay that echos reflected by a hybrid circuit in the subscriber interface (see Chapter 2) of the fixed service would be disturbing. The echo canceller eliminates this echo.

5.5.1.10 Short Message Service Center

GSM provides a paging service that is called short message service. The point-to-point short message service provides a mean of sending messages of a limited size to and from MSs. A *short message service center* (SMSC) acts as a store and forward center for these short messages. The service center is not standardized as a part of a PLMN, but the GSM network has to support the transfer of short messages between SMSCs and the MSs.

5.5.1.11 Operation and Maintenance Center

The OMC is a network management system for the remote operation and maintenance of a GSM network. The alarms of GSM network elements and traffic measurement reports are collected there. The connections to the network elements are arranged by, for example, X.25 or CCS connections. The operation and maintenance system handles features related to system security and network configuration updates.

5.5.1.12 Interfaces Inside GSM/PCS-1900

The interface between the MSC and BSC is as called an A-interface. It is standardized and BSS and MSC from different vendors are compatible. Speech is PCM-coded (see Chapter 3) at this interface. Another important interface is the so-called Abis-interface between BTS and BSC. At this interface speech is GSM-coded, which requires less transmission capacity than the PCM coding. The Abis-interface is not completely standardized and, as a consequence, both BTSs and BSC have to be purchased from the same manufacturer.

5.5.2 Physical Channels

The multiple access scheme used in GSM/PCS1900 utilizes two access methods: TDMA and FDMA. Up to eight users may share one of the 200-kHz frequency channels because it is divided into eight timeslots.

5.5.2.1 FDMA and TDMA

A basic concept of GSM/PCS-1900 transmission on a radio path is that the unit of transmission is a series of about a hundred modulated bits. This is called a burst; see Figure 5.9. These bursts are sent in time and frequency windows called slots. The central frequencies of the slots are positioned every 200 kHz (FDMA) within the system frequency band and they occur every 0.577 ms (TDMA). All timeslots of different frequencies in a given cell are controlled by the synchronization broadcast from the BTS transmitted in the common control channel of that cell.

5.5.2.2 Separation of Transmission Directions in Time and in Frequency

For bidirectional user channels the two directions are related by the fixed separation of frequencies and time. The fixed frequency gap between transmission directions is called the "duplex separation" and it is 45 MHz (in the 900-MHz band) and 95 MHz (1800-MHz band). The separation in time is three timeslots, as shown in Figure 5.10. This principle makes the implementation of mobile equipment efficient because there is no need to transmit and receive simultaneously. Two bursts after the reception on downlink or forward frequency, the mobile equipment sends on the uplink or reverse frequency; as shown in Figure 5.10.

5.5.3 Logical Channels

The physical channels at the GSM radio interface are divided into logical channels. They fall into two main categories, dedicated channels and common control channels. There are many different kinds of logical channels and the distinction between them is based on the purpose and the information transmitted via a channel. Each of these logical channels is mapped into one physical channel defined as a slot and transmitted as a burst.

5.5.3.1 Traffic Channel and Associated Slow-Rate Channel

When the call is connected, there are two channels on the radio path dedicated to it and they are the *traffic channel* (TCH) and the *slow associated control*

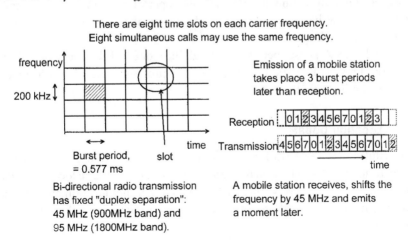

Figure 5.10 Multiple access scheme of GSM and DCS1800.

channel (SACCH). The SACCH is used, for example, to transmit power control information to the MS and measuring results from the MSs to the network. These two channels belong to the dedicated channels because they are allocated for one user.

5.5.3.2 Common Control Channels

There are several logical common channels in each cell. These altogether typically occupy one fixed timeslot at a fixed frequency. The common control channel in the downlink direction, from network to MS, is used to transmit, for example, the following information from the network to the MSs:

- Synchronization information of frequency and timeslots;
- Common channels that are used by neighboring cells;
- Location area and network identification;
- Paging messages for incoming calls and allocated channel.

In the uplink direction, from MSs to the network, the common control channel is used, for example, for call request messages from the MSs.

5.6 Operation of the GSM/PCS-1900 Network

In this section we will introduce an example of the operation of a cellular network. In order to do this we illustrate the GSM network with a few simplified examples. They show how location update is performed, how a mobile call is established, how handover is performed, and what the security functions of the GSM network are.

Each GSM subscriber is registered into one HLR of his own home network. This HLR is the central point that provides subscriber information wherever he is presently located.

5.6.1 Location Update

The cellular mobile network has to be aware of the location of its subscribers all the time in order to be able to route the incoming calls. The location update procedure takes place every time a MS moves to another location area or when a user switches her telephone on in a different location area than where she was located previously.

The geographical position of a GSM mobile is known at the accuracy of a *location area* (LA), which typically consists of a number of cells (BTSs) connected to one BSC. When an incoming call to a mobile subscriber arrives, it is paged through all the cells belonging to the LA where this specific subscriber is known to be.

The MS is responsible for location updates and performs it in idle mode, that is, when a call is not connected.

The MS surveys the radio environment constantly and, when it detects that it could be served best in a new LA, performs a normal location update procedure to change the location information in its present VLR and in the HLR. We say that the MS has roamed to another LA. In dedicated mode, during a call, the procedure called handover, which we will discuss later, may be required. If the LA is changed during a call, the location update takes place after the call is cleared.

Location update may take place inside one network when the LA is changed or between different networks that may be located in different countries. The latter case requires a roaming agreement between the operators to allow a subscriber to use the another network in addition to her home network. Figure 5.11 illustrates the location update procedure that occurs when a MS is switched on in another network Y in another country. This example assumes that the MS has been switched off in the home network, network X, and the network operators of networks Y and X have a roaming agreement that allows cellular subscribers to use the services of another network.

For location updates the following main operations are carried out (see Figure 5.11):

1. When the MS has roamed to another LA, it scans the common control channels. When it finds a common control channel, it detects the LA code. If the LA code transmitted by the network is different from that which the MS has stored previously, it requests a location update from the network.

2. The MSC/VLR receives the global identity code of the mobile. With the help of this global identity information the MSC/VLR knows in which country the home network of this mobile is found. The MSC/VLR sends a signaling message via the international CCS7 signaling network toward the home country of this cellular subscriber. The message includes country code, network code, and subscriber identity. The message also includes the address of this new VLR to inform the HLR about the new location of the MS.

3. When the HLR receives the message it requests the former "old" VLR, where this subscriber was previously located, to remove information of this subscriber.

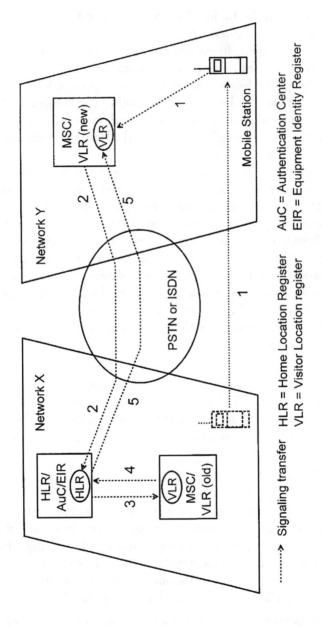

Figure 5.11 Location update in GSM/PCS-1900 network.

4. The VLR (old) acknowledges and removes the subscriber information from its data base.

5. The HLR updates the location information and sends the subscriber information to the new VLR.

The new VLR now knows the security codes of the roamed subscriber and what services this subscriber is entitled.

5.6.2 Mobile Call

Figure 5.12 illustrates how the GSM network routes a call to a subscriber that has roamed to another network. We assume here that both the calling and called subscribers are originally registered in the same home network, network X. The called subscriber B has traveled to another network Y and switched on her MS. Then the location update, which we illustrated in the previous section, has taken place. Then mobile user A calls MS B from the home network.

We can identify the following main phases when the call is established from the home network to a GSM subscriber located in another network; see Figure 5.12.

1. The MS A initiates a call to MS B, which is presently located in another network. The call connection request and other signaling information are transmitted via the radio path and BSS to the MSC. The telephone number of subscriber B (or a name converted into digits by the mobile equipment) is transmitted to the MSC/VLR.

2. The MSC recognizes mobile B (in this example) as a subscriber of its own network and requests the roaming number from the HLR of subscriber B. The roaming number is a temporary telephone number that is used to establish a call via a PSTN.

3. The HLR of subscriber B knows the identification of the "visited" VLR where mobile B is presently located. When mobile B was switched on, network Y informed the HLR under which VLR MS B is now located (location update). The HLR builds up a signaling message that includes the identification of the called subscriber B together with the address of the visited MSC/VLR.

4. The HLR requests the visited VLR to provide a roaming number.

5. The MSC/VLR of network Y has a pool of roaming numbers that look out as ordinary telephone numbers of that country. The visited

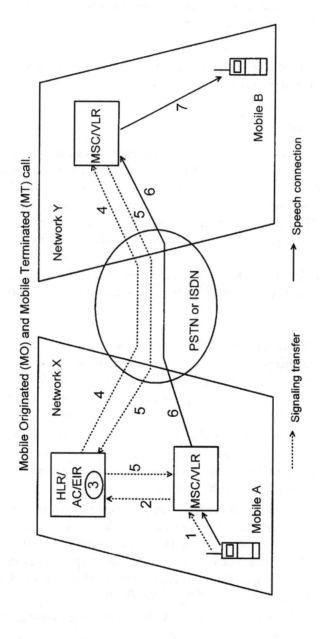

Figure 5.12 Mobile call in GSM/PCS-1900 network.

MSC/VLR then allocates one roaming number for subscriber B and sends it to the HLR, which forwards it to the MSC/VLR in network X.

6. The MSC/VLR of network X routes the call toward the MSC/VLR in network Y using the roaming number for dialing digits, just as in the case of an ordinary telephone call.

7. When the MSC/VLR in network Y receives the call identified by a previously allocated roaming number, it associates this with subscriber B and initiates paging toward MS B. The roaming number is then released for reuse.

The MSC/VLR of network X in the figure acts as a *Gateway-MSC* (GMSC). There is always at least one GMSC in each individual GSM network. The GMSC is a signaling interface point to other networks and is able, for example, to route signaling messages toward the right HLR inside the network.

The telephone call to a roamed subscriber is presently always connected via the GMSC of the home network. In the future it may be possible to connect calls directly. The main concept to be agreed upon between operators is the charging principle in these cases.

5.6.3 Handover or Handoff

The main reason to perform handover is to maintain call connection regardless of the movement of the MS over cell boundaries. The structure of a GSM network requires the possibility to execute handovers at four levels, as shown in Figure 5.13.

The BSC is responsible for handover because it occurs most often between two cells under one BSC. The handover process should be as quick as possible so that communication is not disturbed. In order to perform it quickly the BSC collects measurement data from MSs and BTSs, processes it, and updates ordered candidate cell lists for handover for all the MSs that have an ongoing call.

Handover is most often necessary between BTSs of neighboring cells (the case (b) in Figure 5.13), which are controlled by the same BSC. The BSC controls the handover and performs the channel switch from an "old" cell to a "new" cell. Sometimes there may be a need to switch communication from one channel to another in the same BTS (case (a) in Figure 5.13). This may be necessary because of high interference. Also this handover is controlled and performed by the BSC.

The inter-BSC handover (case (c) in Figure 5.13) occurs if an MS moves to a neighboring cell that is controlled by a different BSC. Now the BSC

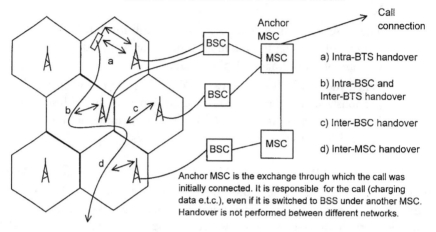

Handover is needed to keep communication quality of the moving MS good enough when a call is connected (mobile station is in so-called dedicated mode). There are four different cases of handover in GSM.

a) Intra-BTS handover

b) Intra-BSC and Inter-BTS handover

c) Inter-BSC handover

d) Inter-MSC handover

Anchor MSC is the exchange through which the call was initially connected. It is responsible for the call (charging data e.t.c.), even if it is switched to BSS under another MSC. Handover is not performed between different networks.

Figure 5.13 The four different cases of handover.

cannot perform switching. Instead, it has to request the MSC to execute the handover switching to the target cell. When a new connection is established from the MSC to the new BTS, the MSC performs the switching.

When the neighbor cell is located under a different MSC, an inter-MSC handover may be required. Then the BSC requests the anchor MSC to establish the connection with the help of a new MSC to the new cell. The anchor MSC performs the switching. The anchor MSC is the exchange via which the call was originally connected. The anchor MSC controls the call until it is cleared even though it may use other MSCs to maintain the call through handovers.

Handover can also be executed to arrange traffic between cells to avoid unsuccessful calls due to geographically uneven load. In this case the calls of the MSs that are located close to the border of a loaded cell are switched to a neighbor cell that has more free capacity.

As an example we now look at how the most complicated handover, inter-MSC handover, is performed. The handover procedure between two MSCs, shown in Figure 5.14, includes the following main phases.

1. The MS moves across the cell border and the BSC (old) decides to initiate handover to another cell (new). The decision is based on the measurement information sent by the MS and by the BTS. The measurement information includes, in addition to information about the traffic channel in use, identifications of neighbor cells and the measurement results of them. The MS continuously measures the

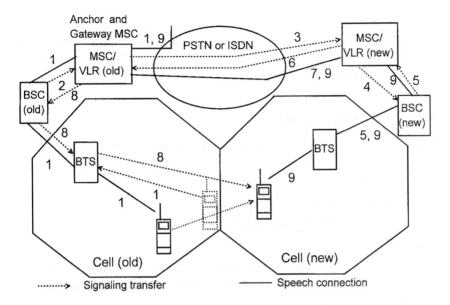

Figure 5.14 Handover or handoff between two MSCs.

common channel of each neighbor cell in addition to the traffic channel in use for the call.

2. The BSC (old) requests the MSC (old) to begin handover preparation to the new cell. The MSC (old) recognizes that the proposed cell (new) is connected to another MSC.

3. The MSC (old) requests a handover number from MSC (new). The handover number is a temporary telephone number that is used to establish a connection via the PLMN or the PSTN for the handover.

4. The new MSC requests an allocation of a traffic channel from the BSC (new).

5. The BSC (new) allocates a free traffic channel and informs the MSC (new).

6. The MSC (new) allocates a handover number and sends it to the MSC (old).

7. The MSC (old) routes a call through the PSTN/ISDN toward the MSC (new) using the handover number as dialed digits.

8. When the routing is complete, the new MSC/VLR commands via the MSC (old) the MS to switch to the new traffic channel (frequency and timeslot) of the new cell.

9. The MSC (new) connects the speech path through the reserved channels in the new cell. Notice that the call is still controlled by the old
MSC, which is called an anchor MSC (e.g., charging records are
produced by it).

5.6.4 The Security Functions of GSM

In the GSM special attention is paid to the security aspects, that is:

- Security against forgery and theft;
- Security of speech and data transmission;
- Security of the subscriber's identity.

Using a radio transmission makes the PLMNs particularly sensitive to
the misuse of resources by unauthorized persons and the eavesdropping of
information exchanged on the radio path.

For security functions the AuC delivers random numbers and precalculated keys for authentication and ciphering to the HLR. It then sends them
with other subscriber information to the VLR, when a location update is
performed. We now review the four most important security functions of
GSM. They are shown in Figure 5.15.

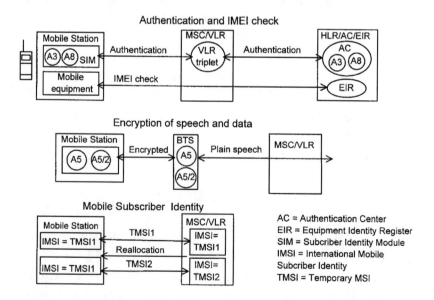

Figure 5.15 The security functions of GSM.

5.6.4.1 Authentication

The principle of authentication is to compare the subscriber authentication key Ki in the AuC and in the SIM without ever sending the Ki on the radio path. For authentication the network sends a random number to the mobile at the beginning of each call. The mobile then uses an algorithm A3 to process a response that is dependent on the random number as well as on the secret subscriber-specific key Ki stored in the SIM. The AuC has also calculated this response and, if they match, call connection is allowed.

5.6.4.2 IMEI Check

The *international mobile equipment identity* (IMEI) check procedure is used to ensure that the mobile equipment does not belong to the black list where the EIR stores the serial numbers of the stolen mobiles. The IMEI is a manufacturer-specific code that is stored in each mobile equipment when it is manufactured.

5.6.4.3 Encryption of Speech and Data

Speech and data are encrypted before forwarding the radio or air-interface. The two algorithms A5 and A5/2 are defined for encryption, and one of these is usually in use. An encryption algorithm uses the ciphering key that is calculated by the AuC and by the SIM. The ciphering key depends on the subscriber-specific key together with the random number that is given to the MS at the beginning of each call and used for authentication as well.

5.6.4.4 Mobile Subscriber Identity

The MS is normally addressed over the air-interface using the *temporary mobile subscriber identity* (TMSI), which is allocated for each mobile located inside a LA. The global identity of the mobile, the *international mobile subscriber identity* (IMSI) is very seldom sent over the air-interface to prevent eavesdropping devices from using it as trigger information. A new TMSI is allocated for each call when communication is in ciphered mode.

We used GSM/PCS-1900 here as an example of a modern digital cellular radio system and illustrated the functionality of the network only with a few simplified examples. A more detailed study of the cellular network functionality is beyond the scope of this book. More comprehensive descriptions about the operation of GSM are given in [1,2].

5.7 Problems and Review Questions

Problem 5.1: What are the main advantages of cellular systems compared with the old generation radio telephone systems that did not utilize a cellular network structure?

Problem 5.2: An analog radio telephone network has a frequency band of 100 (bidirectional) FDMA channels. The network covers a 50- by 50-km urban area. Give the maximum number of simultaneous calls in the network if: (a) only one BS is in use; (b) the network is upgraded to a cellular network with a cell size of 10- by 10-km and reuse ratio is 1:9 (each channel is used again in every ninth cell); (c) cell size is reduced to 1- by 1-km; and (d) cell size is reduced further to 0.35- by 0.35-km (that was equal to the minimum size of cells GSM in phase 1). For simplicity we may assume here that the cells are rectangular and all the channels are used as traffic channels. Hint: Divide all channels of the network between a cell cluster (group) of nine cells. Then repeat this cluster to cover the geographical area of the network.

Problem 5.3: What are the two main types of channels used in each cell of a cellular mobile system?

Problem 5.4: What is handover? Explain the principle of how it is carried out in a cellular network.

Problem 5.5: How does the cellular network route an incoming call to a subscriber located anywhere in the network or even in a different country? What are the roles of HLR and VLR in the routing of an incoming call?

Problem 5.6: Explain the main phases that occur in the radio interface of a cell when an outgoing call is requested. Explain also what happens when an incoming call is received by an MS in a cellular network.

Problem 5.7: Explain the applications of cordless telephones. How do cordless systems basically differ from cellular systems?

Problem 5.8: Explain the structure of a GSM network. What are the network elements and what are their roles in the operation of GSM?

Problem 5.9: Explain the multiple access method of GSM.

Problem 5.10: Explain how a location update is performed in GSM. What triggers it and what happens after that?

Problem 5.11: Explain how a call is routed from a GSM MS to another MS of the same network.

Problem 5.12: Explain how the handover is performed in the GSM network.

References

[1]   Redl, M. S., M. K. Weber, and M. W. Oliphant, *An Introduction to GSM,* Norwood, MA: Artech House, 1995.

[2]   Mouly, M., and M. B. Pautet, *The GSM System for Mobile Communications,* Paris: Michel Mouly and Marie-Bernadette Pautet, 1992.

6

Data Communications

In this chapter we clarify some key terms that we need in order to describe a certain data communication principle or a system. Then we introduce the concept of data communication protocols trying to get a concrete touch on layered data communication protocols and the reason why we define data communication architectures with the help of the protocol layers. Then we describe the characteristics of various data communication systems and the requirements of various services. In the latter half of this chapter we describe different systems for local and wide area data communications.

6.1 Principles of Data Communications

The first data communication system was the telegraph. It was invented more than a hundred years ago. The letters to be transmitted were converted into a code called the Morse Code. The codes were transmitted as pulses along a wire or as radio frequency bursts in the case of wireless telegraph. Then development of data communications was very slow, but during the last twenty years data communications have expanded rapidly as computers have become tools for everyone.

6.1.1 Computer Communications

Modern computers manipulate bits, binary symbols, of electrical energy. When a computer communicates with another computer it sends these bits along a cable between them. This is relatively easy if the computers are within the same room or a building. If the distance is longer there is a need to use a telecommunications network that provides an end-to-end communication channel. There are several different variations of how data communication can be arranged; we will discuss some of these in the following subsections.

6.1.2 Serial and Parallel Data Communication

In a transmission network only one channel is usually allocated for one end-to-end connection in each direction. Let us use a simple *American Standard Code for Information Interchange* (ASCII) as an example of the data source. We press keys on the keyboard and each key stroke generates a seven- or eight-bit binary word corresponding to the letter or number of the key pressed. If we have only one channel available, we have to send bits of this word in turn to the channel and in this case we talk about serial data transmission; see Figure 6.1.

In serial transmission we need only one channel but have to use a line code to insert timing information into the data stream. This synchronization information enables the receiver to know when it has to detect each individual received bit. How we implement this depends on whether we use asynchronous or synchronous transmission mode. We describe them in Section 6.1.3.

If a computer needs to communicate with, for example, a printer in the same room, parallel communication is often used. A special cable with several wires is provided between the computer and the printer and all eight bits of a data word, corresponding to one character, are transferred at the same time in parallel over the cable. Parallel data transmission is much quicker than serial,

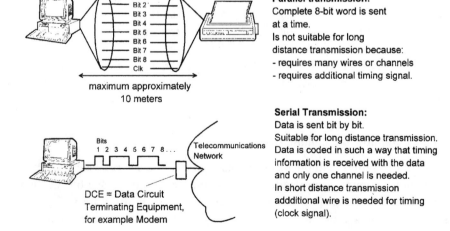

Figure 6.1 Serial and parallel transmission.

but we can typically use it only over short distances. The maximum is usually in the order of 10m.

Communicating terminal devices in data communication are called *data terminal equipment* (DTE) and the equipment that terminates the transmission channel that goes through the network is called *data circuit-terminating equipment* (DCE). A modem that we use for data transmission over a telephone network is a typical example of DCE. There are many different interface specifications between DTE and DCE, and the most common standards are defined by ITU-T and *Electronic Industries Association* (EIA) in the United States. One of the most common data interfaces is called V.24/V.28 of ITU-T; it corresponds to the EIA standard RS-232-C.

6.1.3 Asynchronous and Synchronous Data Transmission

Over longer distances we use serial transmission either in asynchronous or synchronous transmission mode. Serial transmission requires that the timing information for the receiver is transmitted together with the data so that a separate clock signal is not required.

In asynchronous transmission only a small number of bits are transmitted at a time, usually eight bits that correspond to one character of an ASCII terminal. In the beginning of each block of eight bits of data, a "start" bit is sent to indicate to the receiving computer that it should prepare to receive eight bits of data; see Figure 6.2. For synchronization the receiver has to know the data rate so that when it detects the start bit it is able to receive the few following bits. After these data bits a stop bit is sent. The next block of data

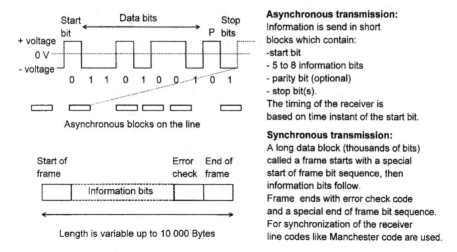

Figure 6.2 Asynchronous and synchronous transmission.

is synchronized independently with the help of a new start bit preceding the data bits.

In asynchronous transmission a simple error-detecting scheme called parity is used. We may use even or odd parity error check. If even parity is used, the total number of "1" bits in the block, including data bits and the parity bit, is set to be even with the help of the parity bit. In the case of odd parity, the parity bit is set to "1" or "0" so that the total number of "1" bits in the block is odd. To detect possible transmission errors the receiver checks if the received number of "1" bits is even or odd depending on the parity agreed. We will see later that this parity check method is a simple example of data link layer protocols.

Asynchronous transmission is used for the transmission of ASCII characters in conventional terminal-mainframe computer communications. For larger information blocks it is used in some file transfer protocols such as KERMIT, X-LINK, and YMODEM. In these protocols special "start of block" characters are sent at the beginning. Then information follows as asynchronous words and in the end special "end of block" characters are sent.

Synchronous transmission is a more modern principle for transmitting a large amount of information in a frame; see Figure 6.2. Each frame starts with a special bit sequence and the frame may contain more than a thousand bytes of information. Each frame also contains error control words and an end-of-frame sequence. The receiver uses the error control section of the frame to detect if errors have occurred in transmission. The most common detection method is called *cyclic redundancy check* (CRC). It is much more reliable than the parity check method that we discussed previously. In the case of errors the transmitter retransmits the frame in error. In the most common protocols the receiver sends an acknowledgment to the transmitter in the other transmission direction for each received error-free frame. If errors have occurred, the frame is not acknowledged and the transmitter sends it again after a certain time constant.

In asynchronous transmission the start bit provided the required timing information for each byte of data. In synchronous transmission data blocks are not divided into separate bytes and it is required that the bit timing information be inserted into the data stream itself with the help of line coding. As an example, LANs use the Manchester line code that we described in Chapter 4.

6.2 Data Communication Protocols

The computers that communicate have to understand each other. They have to speak the same "language." This common language is defined as a data communication protocol. A detailed protocol specification that enables two different systems to communicate includes many communication rules such as how the letter "A" is presented as a binary word and what is the voltage of bit "1".

As we see, there are many specifications needed to enable data communication between systems. A standard called *Open Systems Interconnection* (OSI) gives guidelines on how this complicated set of communication rules is divided into smaller groups of rules and functions that are called layers. This helps us concentrate on one group of functions (= protocol of a certain layer) at a time and we do not have to care about the functions for which other layers are responsible. For example, if we are specifying the error detection code that belongs to the data link layer of OSI, we need not worry about the power levels of optical transmission or the shape of electrical pulses in the coaxial cable. These issues are the problems associated with the lowest layer, the physical layer in OSI.

In the next section we will review the OSI reference model that was standardized by the *International Standards Organization* (ISO) and try to clarify the importance of the principle of the layered structure in data communications.

6.2.1 Protocol Hierarchies

To reduce the design complexity of computer communications hardware and software, the needed functionality is organized as a series of layers, each built upon its predecessor; see Figure 6.3. There are many proprietary protocols in use in addition to the available international standards. All of them use some form of layering. The number of layers, the name of each layer, the contents of each layer, and the function of each layer may differ from network to network.

In all networks, the purpose of each layer is to offer certain services to the higher layers, shielding those layers from the details of how the provided services are actually implemented.

6.2.1.1 Protocol

Each layer in one machine carries on a conversation with the corresponding layer in another machine; see Figure 6.3. The rules and conventions used in this conversation are collectively known as the protocol of this layer. We can say that the protocol specifies the format and meaning of the information that a layer sends down to the layer below. This information is received and understood by the corresponding layer at the other end if exactly the same protocol specification is implemented there.

With the help of its protocol each layer below provides services to the layer above it. The service specification is sometimes separate from the protocol

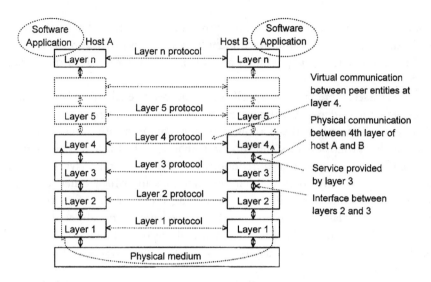

Figure 6.3 Protocol hierarchy.

specification. We could say that service specifies what the layer looks like from the point of view of the above layer. For example, if a layer provides data transmission with or without error detection, the layer above may select which one it wants to use. How they are implemented in the layer is specified in the protocol specification.

The interfaces between layers are defined to be as simple and clear as possible and each layer performs a specific collection of well-understood functions.

6.2.1.2 Protocol Stack

The set of layers and their specified protocols is known as a protocol stack. For successful communication both computers have to use exactly the same protocol stack where each layer complies with the same detailed standard.

6.2.2 The Purpose and Value of Layering

The purpose of each layer is to provide certain services to the higher layers, shielding those layers from the details of how the provided services are actually implemented. Without this abstraction technique it would be difficult to partition the design of communication hardware and software into smaller manageable design problems, namely the design of individual layers.

This also makes it possible to replace one layer with a new implementation without affecting other layers. Consider, for example, a LAN where the same software applications use both Token Ring and Ethernet network technologies. We illustrate the fundamental idea of the layered protocol structure in the following subsections.

6.2.2.1 An Analogy to Explain the Idea of Multilayer Communication

Imagine that two philosophers (at layer 3), one in Egypt and one in the Philippines, want to communicate remotely (see Figure 6.4). The philosophers have their own jargon specific for their profession and only another philosopher can understand it completely. This corresponds to the protocol of layer 3.

Since these philosophers have no common language, they each need a translator (at layer 2). To establish a communication channel each translator contacts an engineer (at layer 1). When the Egyptian philosopher wishes to discuss with another philosopher he passes the message across the 3-2 interface to his translator at layer 2 who uses the language that he has agreed previously with the other end. The translators use their best common language, which may be English, and this agreed common language of the translators corresponds to the layer 2 protocol.

The translator then gives the message to layer 1 for transmission. Engineers at layer 1 may use any channel they have agreed in advance. This physical communication may use a telephone network, a leased satellite channel, a computer network, or some other means. This engineer and the communication channel arranged by him correspond to the layer 1 protocol.

The purpose of layering is to offer certain services to the higher layers, shielding those layers from the details of how the offered services are actually implemented.

Additional advantages of standardized layers are:
- Easier engineering (complicated system is divided into managable pieces).
- More efficient to develop further (replacement of one layer).
- Definition of layer responsibilities help standardization of new functions.
- Certain functions belong to a protocol of a certain layer.

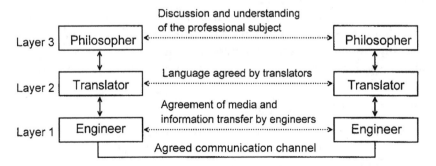

Figure 6.4 Purpose and value of layering.

When a message arrives to the Philippines, it is translated by the translator (layer 2) at that end and passed to the receiving philosopher. Let us imagine that these English speaking translators are replaced by others, for example, because of a lunch break. These new speaking translators notice that French is a better common language for them and they agree to use that. The service provided by layer 2 remains the same and the philosophers do not notice that the protocol of one lower layer is completely changed.

In the same way engineers may change the channel in use and upper layers may not notice and do not even care how communication is arranged as long as the quality of service is acceptable. Note that each protocol layer is completely independent of the other layers and higher layers do not have to care about how communication is actually arranged by the lower layers, that is, what protocol they use.

6.2.3 Open Systems Interconnection

Back in the late 1970s the ISO began to work on a framework for a computer network architecture known as the OSI Reference Model. The purpose of this model was to eliminate incompatibilities between computer systems.

In 1982, ISO published the document ISO 7498 as a draft international standard. This document is just a framework about how communication protocols should be designed, not a detailed specification needed for compatibility. CCITT/ITU-T published it as recommendation X.200.

OSI was originally designed for computer communications. Today data and voice are not necessarily separated into different networks. At many times the network does not know and is not interested in what information the transmitted data contains. ISO and ITU-T specify all new networks and systems according to the layering principle of OSI. However, there are some worldwide systems that are not designed according to OSI and the most important of them is the Internet. The Internet is based on standards that are openly available but not approved by ISO or ITU-T.

The name OSI comes from the goal to make systems open for communication with other systems. Any manufacturer is free to use these "open" specifications. Anyhow, many data communication systems are still proprietary systems and their specifications are the property of one vendor, so they are not available to others.

6.2.3.1 The OSI Reference Model

In the OSI model communication is divided into seven layers; see Figure 6.5. The OSI reference model just tells what each layer should contain, but it does not specify the exact services and protocols. The detailed specification of each layer is published as a separate international standard.

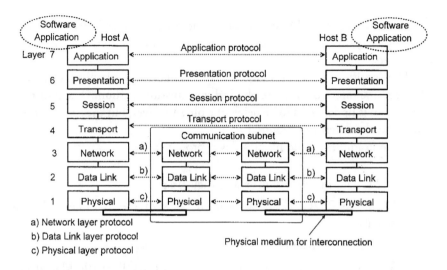

Figure 6.5 The OSI reference model.

Note that the layers below the transport layer take care about the data transmission through the network from node A to node B. The transport layer is the lowest end-to-end layer and uses the network to implement the service for the session layer.

When we look at what kind of functions each layer performs we will notice that the more we look at the lower layer, the more we see functions related to the network technology used for the actual data transmission. The more we look at the upper layers, the more we see common functions available for software applications running in hosts. As we see in Figure 6.5, layers 4 to 7 are all implemented only in the communicating end machines. We do not need layers 4 to 7 at all for actual end-to-end data transmission; this is done by the layers 1 to 3. The only purpose of the uppermost layers is to help software applications, and to do this they provide more sophisticated services than just a stream of data. This stream of data provided by the network layer may contain some errors. Each application software designer should design an error recovery scheme in his application if this service is not provided by the transport layer protocol.

6.2.3.2 Physical Layer

The physical layer is concerned with transmitting bits over a communication channel. The main design issue is to make sure that when one side sends a 1 bit, it is received by the other side as a 1 bit rather than as a 0 bit. Typical specifications of the physical layer include the duration of a bit in microseconds, the number of volts used to represent a 1 and a 0, the number of pins, and the connector type used. Physical layers of the systems are designed to operate practically error-free. If errors occur, the consequential actions are left to the upper layers.

The specifications of the physical layer deal with mechanical, electrical, and procedural interfaces and the physical transmission medium. The transmission medium is understood to be below the physical layer, but the specifications include the characteristics required by it.

Some examples of the physical layer protocols are:

- V.24, data interface signals and their usage;
- V.28, RS 232C, electrical characteristics of the asymmetrical data signals;
- IS 2110, specification of a physical connector;
- V.35, data transmission at 48 kbit/s using the 60- to 108-kHz group band; V.35 is often used to refer only to the high data-rate physical interface specified in the V.35 recommendation;
- X.21, interface for circuit-switched data networks, which includes electrical and procedural specifications related to the establishment of communication;
- I.430 (IS 8877), ISDN basic rate user interface;
- ANSI 9314, specification of optical interface for wideband data network called FDDI.

The first three specifications widely used are often simply called V.24/28 or RS 232C interface.

6.2.3.3 Data Link Layer

The data link layer builds up the frames, sends them to the channel via the physical layer, receives frames, checks if these frames are error-free, and delivers error-free frames to the network layer. The data link layer at the receiving end sends acknowledgment of error-free frames to the transmitting end. The transmitter may retransmit the frame if no acknowledgment is received within a certain time period.

The ISO specifies the data link layer for LANs and divides the specifications into two sublayers:

- *Medium access* (MAC) sublayer;
- *Logical link control* (LLC) sublayer.

This division is necessary for LANs because of the complexity of the data link layer in this kind of application. In LANs computers are connected to the same cable and share the transmission capacity of a broadcast (multiaccess or random access) channel. The MAC sublayer is concerned with functions dependent on the network hardware. The two popular examples of different LAN technologies are the *carrier sense multiple access/carrier detect* (CSMA/CD): "Ethernet" and Token Ring. The LLC considers most of the data integrity aspects, such as retransmission and acknowledgments. In the case of a simpler point-to-point link there is no need for a separate MAC layer and one data link layer protocol specification may cover the whole data link layer.

In a LAN each computer has its own MAC address (hardware address). This address is used to identify the source and destination of each frame in the broadcast channel. With the help of this address computers can have a point-to-point connection via a broadcast channel that is shared by many other point-to-point connections. Note that this address is used only inside a LAN, it is not transmitted to other networks.

Some examples of data link layer protocols are:

- IS 3309, *high-level data link control* (HDLC);
- Q.921, LAP-D, ISDN layer 2, HDLC-based data link layer protocol;
- IEEE 802.3 = IS 8802/3, MAC layer of CSMA/CD ("Ethernet") LAN;
- IEEE 802.2 = IS 8802/2, LLC of multiple access LANs. For a complete LAN data link layer both 8802/2 and 8802/3 are needed.

6.2.3.4 Network Layer

The layers below the network layer are interested only in the point-to-point connections between two nodes. The network layer has some knowledge about the structure of the network and, together with the network layers of the other nodes it services, that packets are routed through the network to the destination. Each node has its own (network layer) global address.

A key issue is the determination of how packets are routed from the source to the destination. Routes can be based on static tables at the network layer that are rarely changed, or they can be dynamic to reflect the current network load.

The hosts connected to the network are autonomously sending packets when they wish. They usually are not informed about the traffic density of other hosts or network connections. If many hosts happen to be active at the same time, too many packets are transmitted and, hence, have the potential to get in the way of each other and form bottlenecks inside the network. The control of such congestion also belongs to the network layer.

In public data networks the accounting function is often built into the network layer. The software in the network layer must count how many packets or characters are sent by each customer in order to produce the charging information.

In an isolated broadcast network (such as Ethernet) routing is so simple that the network layer would not be needed at all. MAC addresses could identify the hosts. However, if and when these networks are connected to other networks, network addresses are mandatory. Note that the MAC-addresses used in the data link layer have no importance outside one LAN.

Some examples of network layer protocols are:

- X.121, addressing of digital networks;
- IS 8208, X.25 packet layer;
- Q.931, I.451, ISDN D-channel, layer 3;
- *Internet Protocol* (IP) of the Internet; not approved by ISO but it performs basically the same functions as network layer protocols of OSI.

6.2.3.5 Transport Layer

The transport layer is the first true end-to-end layer. The protocols of hosts from the transport layer upward use the network as a point-to-point connection for communication. The source message may be split up by the network layers on the way and the destination session layer may be the first point where the pieces, belonging to the same message, meet again.

The transport layer acts as an interface layer between network connection-oriented lower layers and application service-oriented upper layers. Its responsibility is often to check that end-to-end transmission is error-free and packets are not lost on the way. For this it may include end-to-end acknowledgment and retransmission procedures.

The transport layer usually provides two basic service classes to the session layer:

- Transport of isolated messages and datagrams through the network. Transmitted messages may arrive in different order and errors may occur. Examples include the *User Datagram Protocol* (UDP) of the Internet (actually this does not belong to OSI-protocols) and *Transport Protocol, class one* (TP1) of OSI (IS 9072).
- Error-free point-to-point channel delivers messages in the same order that they were sent. Examples of these are *Transmission Control Protocol*

(TCP) of the Internet (not included in OSI protocol standards) and TP4 of OSI (IS 8072/8073).

6.2.3.6 Session Layer

The transport layer ensures that end-to-end transmission between computers is successful. Actually, the task of communication is accomplished by the four layers below the session layer. The three uppermost layers are not needed for data transmission, but they help make applications compatible so that the application programs running in computers understand each other.

The session layer allows users on different machines to establish sessions between them. It can be used, for example, to allow a user to log into a remote time-sharing system or to transfer a file between two computers.

A session layer allows ordinary data transport, as does the transport layer, but it also provides some enhanced services useful for some applications. One of these services is to manage dialogue control. Sessions can allow traffic in both directions at the same time or in only one direction at a time. If traffic is allowed in only one way at a time, the session layer can help by keeping track of whose turn it is. The session layer also provides a token management function and, with the help of this, only the one holding a token may perform a critical operation.

Another service by the session layer is to support successful transmission of large files. Without this service a single error might destroy the whole file that then should be retransmitted. To eliminate this problem, the session layer provides a way to insert checkpoints into the data stream so that after a crash only the data after the last checkpoint has to be repeated.

An example of the session layer standards is the International Standard IS 8326/8327 (X.215/225 of ITU-T) that defines the connection-oriented session layer service and protocol. By connection-oriented we mean that a connection is established between communicating entities before data transmission may start.

6.2.3.7 Presentation Layer

As we saw, the lower layers mainly deal with the orderly transfer of bits or data from source to destination. The presentation layer instead is concerned with the format of the transmitted information. Each computer may have its own way of representing data internally, so agreements and conversions are needed to ensure that different computers can understand each other.

It is the job of the presentation layer to encode the structured data from the computer's internal format to a bit stream suitable for transmission. This may require compression, for example. The presentation layer at the other end decodes the compressed data to the required representation at the destination.

The presentation layer helps both computers understand the meaning of the received bit stream in the same way.

Different computers have different internal representations of data. All IBM mainframes use *extended binary-coded decimal interchange code*, (EBCDIC) eight-bit codes as character code; whereas practically all others use ASCII, seven- and eight-bit options. The Intel chips number their bytes from right to left, whereas Motorola chips number theirs from left to right. Since computer manufacturers rarely change these conventions, it is unlikely that any universal standards for internal data representation will ever be adopted.

One solution for ensuring compatibility is to define a standard for the "network representation" of data so that any computer may communicate with another if each of them converts its internal representation to this standardized network format.

Other tasks for the presentation layer are data compression and encryption. Some examples of presentation layer protocols are:

- IS 8824: a standard for the representation of data structures, ASN.1, Abstract Syntax Notation 1 (abstract because it is just a representation). ASN.1 descriptions are quite similar to any high-level programming language including definitions of data structures such as integer and floating point number.
- IS 8825: encoding rules for ASN.1 defining how representations are encoded into a bit stream for transmission.

6.2.3.8 Application Layer

The application layer contains the communication applications that use the services of the lower layers. User applications that perform the tasks for which computers are purchased are not included in the application layer, but they communicate with the help of the application layer protocol. An example of user applications is a word processing application.

Often needed communication applications such as file transfer or an ASCII terminal have been defined as the application layer protocols to serve any user application. Communication applications provide common vendor-independent services for user applications of any vendor. The application layer services are usually available for the programmer as any other services of the operating system in use. With the help of these services software application programmers (designing, e.g., word processing software) do not have to worry about actual data transmission at all. He may use all the services of the protocol stack implemented in his development environment.

One example of application protocols is electronic mail. In addition to a service similar to file transfer, it provides ready-made functions for deleting, sending, and reading mail. The specifications of the application layer define, for example, the format of addresses and message fields.

To distinguish between application programs and the application layer defined by a protocol, let us look at an example of electronic mail. We may have an application running on top of the application layer in our computer. This program may provide nice colors, a user-friendly editor, and separate windows for addresses and messages. It may also provide a user-friendly addressing method; that is, we can give a destination address such as "John" that is converted by the software to the format that the application layer understands. Note that the application layer service provides the required communication services required but we may enhance them with an application software for local purposes.

Some examples of application protocols include:

- X.400, *message handling system* (MHS) of ITU-T; electronic mail;
- IS 8571, *file transfer access and management* (FTAM) of ISO; file transfer protocol of OSI.

Those readers who are not familiar with any data communication protocols may have found this section quite abstract. To make the operation of protocol layers more concrete, we will illustrate in the next section how actual data packets are handled when they are transferred down and up through the protocol stack.

6.2.4 Data Flow Through a Protocol Stack

Let us assume that the user of the source machine performs an action that creates the message, M(A), which is produced by a process running in the application layer (7) in the source machine; see Figure 6.6. This message could be an e-mail that we transmit to another computer through the network. The message is passed from layer 7 to the presentation layer (6). In this example, the presentation layer transforms the message in a certain way (e.g., text compression) and then passes the new message M(P) to the session layer (5). The session layer, in this example, does not modify the message but regulates the data flow to prevent an incoming message from being handed over to the presentation layer while it is busy. Data units given to the lower layers are called protocol data units, for example, *presentation protocol data unit* (PPDU).

In most networks a data packet has a certain maximum length, but usually there is no limit to the size of messages accepted by the transport layer. If the message is very long, the transport layer must break it up into smaller units,

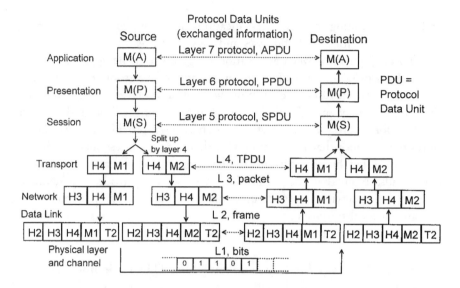

Figure 6.6 Data flow through a protocol stack. (Source: Tananbaum, A., *Computer Networks,* Englewood Cliffs: Prentice-Hall, 1988.)

adding a header to each unit. The header includes control information such as a sequence number. In many networks transmitted units may arrive in a different order than they were transmitted. With the sequence number the transport layer at the destination machine is able to build up the original message by placing the transmitted pieces into the correct order.

The network layer (3) looks up the routing table and decides which of the outgoing lines to use. It attaches its own headers such as the address of the destination network layer and passes the data to data link layer (2). The network layer message is often called a packet.

The data link layer adds a header and a trailer and gives the resulting unit to the physical layer for transmission. The header may include a start-of-frame flag and physical addresses in a LAN. The trailer is needed for error detection and an end-of-frame flag. The message of the data link layer is often called a frame.

The physical layer transmits the bits given by data link layer to the physical media, such as cable. It may convert bits into light pulses for optical fiber cable transmission.

In the receiving computer the message moves layer by layer upward. The header of a layer is stripped off by a corresponding layer at the other end. None of the headers for layers below a certain layer n are passed up to the layer n. Thus, each layer receives the message as it was transmitted by a corresponding layer in the source machine. They act as if they were connected directly, not through the lower layers. For example, when the data link layer of the destination machine has checked through the error detection field in the trailer (T2) that the frame is error-free, the error-check bytes are removed before the data is given to the network layer.

If the reader still feels that the preceding illustration was too abstract and wants to understand the principle of protocols and layers thoroughly, she may study the operation of one protocol stack, for example, TCP/IP, layer by layer. This is the most efficient way to get a concrete grip on protocols; and when one protocol is understood, new protocols are easy to learn. A comprehensive description of TCP/IP is given, for example, in [1].

In the following sections we look at the basic functions and characteristics of various data communication systems.

6.3 Voice Band Modems

The word modem comes from the combination of the two functions, modulation and demodulation. Modulation converts a digital signal into an analog signal for transmission through a channel and demodulation performs the conversion back to the original digital signal. Voice band modems are needed when an analog voice channel of the telephone network is used for data transmission.

6.3.1 Operation of the Voice Band Modems

The frequency band of the voice channel is 300 Hz to 3400 Hz, and the baseband digital information is transferred to this band through of CW modulation. The CW modulation methods used in carrier wave or voice band modems are exactly the same that we used for radio transmission in Chapter 4.

As we know, CW modulation may vary three characteristics of a carrier: amplitude, frequency, or phase. The corresponding basic modulation methods are *amplitude modulation* (AM), *frequency modulation* (FM), and *phase modulation* (PM). All these basic modulation methods are used in the voice band modems.

As we see in Figure 6.7 the only analog section in the connection through a modern telecommunications network is the subscriber line of the LAN. The fastest standardized voice band modems can support data rates up to 33.6 kbit/s. The maximum user data rate is in the order of 30 kbit/s even though the transmission rate inside the PSTN is 64 or 56 kbit/s (data rate of PCM-coded voice channel). More than half of the end-to-end data capacity

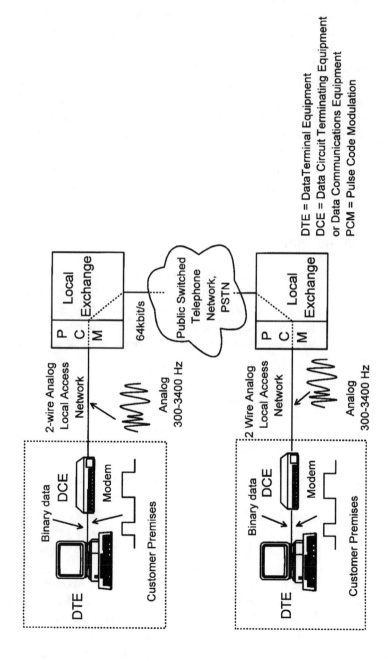

DTE = DataTerminal Equipment
DCE = Data Circuit Terminating Equipment
or Data Communications Equipment
PCM = Pulse Code Modulation

Figure 6.7 Modem link over the PSTN.

is wasted because of analog subscriber lines that perform A/D-D/A conversions at both ends.

New modems with essentially higher capacities will not be standardized because it is expected that ISDN will take over and modems are already quite close to the theoretical maximum capacity of the voice channel. When analog subscriber lines are replaced by ISDN lines, the full capacity of the allocated channel in the network can be utilized. The end-to-end data rate will then be 64 kbit/s (B-channel) and 128 kbit/s (two B-channels).

6.3.2 V-Series Recommendations of ITU-T

The ITU-T (CCITT) defined many standards for modems with a variety of speeds and these recommendations are identified by the letter V and a number attached to it. Modems of different manufacturers work together if they support a compatible V-standard. Many modern modems support previous lower speed standards as well and they are able to adapt their speed on the level supported by the other end. Some examples of modem standards are as follows [2]:

- V.21: 300-bit/s full-duplex (bidirectional transmission); the carrier frequencies at different transmission directions are 1080 Hz and 1750 Hz; FSK is used; binary "1" corresponds to the carrier frequency of the direction in question (1080 kHz or 1750 kHz) minus 100 Hz and binary "0" corresponds to the carrier frequency plus 100 Hz.

- V.22: 600/1200-bit/s full-duplex standard that provides an acceptable dial-up data connection for the transfer of short text messages in both directions. One example of user applications is a remote text-mode terminal. The modulation scheme is PSK with two or four phases; modulation rate is 600 Bauds.

- V.22bis: 2400-bit/s full-duplex; the transmission directions are separated by frequency division. This modem is an update to V.22 in the end of 1980s. The data rate was doubled with four phases and four amplitudes of the carrier; the modulation rate is still 600 Bauds.

- V.23: 1200/75-bit/s modem that transmits 1200 bit/s in one transmission direction and 75 bit/s in another direction. This asymmetrical transmission provides enough capacity to send key strokes from the terminal while transmitting larger amounts of data in the other direction. FSK is used in both directions and 1300 Hz corresponds to "1" and 2100 Hz corresponds to "0" in the 1200-bit/s direction. In the 75-bit/s direction frequencies are 390 Hz as "1" and 450 Hz as "0".

- V.32: 9600-bit/s full-duplex standard that is capable of transmitting 10 pages of text in approximately 30 sec without data compression. Modulation method is QAM, which is a combination of amplitude and phase modulation. The modulation rate is 2400 Bauds, and 16 combinations of carrier phases are used.

- V.32bis: This modem is an enhancement of V.32 with a new modulation scheme QAM. It transmits data at 14.4 kbit/s. This data rate is barely acceptable for modest graphical applications such as the *World Wide Web* (WWW) of the Internet. The modulation method is QAM with 128 different combinations of amplitude and phase of the carrier, and the modulation rate is 2400 Bauds.

- V.34: This standard supports data rates up to 28.8-kbit/s full-duplex over dial-up telephone lines. Error-free operation at this high data rate requires a very clean speech channel. If errors occur too frequently this modem falls back to a lower speed in steps of 2400 bit/s down to 2400 bit/s if necessary.

- V.34+: Enhancement to V.34 with data rate 33.6 kbit/s. Modulation method is QAM and modulation rate is 3200 Bauds, just as in the V.34 standards.

It is most likely that essentially higher data rate voice band modems than V.34+ will not be standardized. Essentially higher data rate service requires an end-to-end digital connection provided by, for example, ISDN.

Note that data transmission with a voice band modem does not require anything else than a modem at the end of the subscriber line and an analog voice band circuit through the network. There are many other means to transmit data that are called modems even when they may not be modems at all, that is, they do not perform modulation and demodulation. One of these systems is the baseband modem that we will look at in Section 6.4. Another modern system is the 56-kbit/s modem that provides high-rate transmission to subscriber premises and lower data rate transmission in the opposite direction. In this system, just as in the case of baseband modem, the high rate data is not carried in an analog telephone channel the same way as speech. Special arrangements are required in the network to support the usage of these modems. The voice band modems that we discussed in this section use the telephone network in exactly the same way as ordinary telephones.

Modems support a variety of functions, not only modulation and demodulation, and some of these additional functions are explained in the following subsections. The interfaces of a voice band modem as well as some of the additional functions are shown in Figure 6.8.

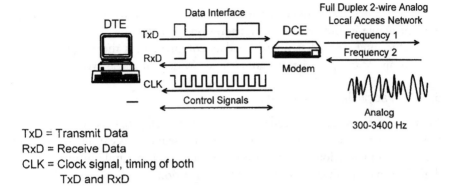

TxD = Transmit Data
RxD = Receive Data
CLK = Clock signal, timing of both
    TxD and RxD

Figure 6.8 A full-duplex modem. Control signals are needed, for example, for flow control that disables the transmission of DTE if the transmission rate via PSTN is too slow, to indicate an incoming call and to command a modem to start dialing.

6.3.3 Additional Functions of Modems

In addition to the basic modem functionality that allows a user to transmit data over an ordinary telephone channel, most modern modems include additional functions, the most important of which are listed in Figure 6.8 and described briefly in the following subsections.

6.3.3.1 Error Control

Errors may occur in the transmission channel, for example, because of the noise in the subscriber line. Many modems send, in addition to data, error-check bits; and with the help of these, they are able to detect and probably correct some bit errors. Both ends have to support the same error control protocol. One international standard for error correction in modems is the V.42 recommendation of the ITU-T.

In addition to error detection and correction in modems, most communication software packages include error recovery functions as well. If a block of data containing errors is received and errors are detected, retransmission is requested by the far-end software.

6.3.3.2 Data Compression

Data compression makes it possible for the transmission rate at the interface between a computer and a modem to be much higher than the actual transmission rate through the network. For example, text can sometimes be compressed by a factor of four or even more. There are several methods for data compression. As a simple example of the compression of text information we can imagine that the most common characters are not transmitted in ASCII form but with very short codes; less frequently needed characters would use longer bit strings. This principle would save some transmission capacity.

One international standard for data compression is the V.42bis recommendation. Many proprietary standards are widely in use as well.

6.3.3.3 Facsimile Transmission

Many modern modems include facsimile functionality that enables a user to send and/or receive faxes without printouts. These modems comply with group 3 fax recommendations of the ITU-T and transmit facsimile information in digital format at 9600 bit/s using FSK modulation.

6.3.3.4 Dial-Up Modems

All modern modems are able to transmit multifrequency signaling tones to the telephone network to establish a connection. Modern modems include signaling functionality similar to that of a telephone so that an external telephone is not needed for dial-up.

6.3.3.5 Data Rates of Voice Band Modems

Modems operate at various data rates over a voice band telephone channel. Many modems support many different data rates and can adapt the transmission data rate to the quality of the channel. In Chapter 4 we saw that the maximum transmission data rate depends on the bandwidth and noise of the channel. If the S/N is degraded (noise level increases), the data rate has to be decreased to keep the transmission error rate low enough. Modems are also able to adapt their data rate and error correction scheme to the capability of the other end. In order to do this they exchange data control sequences while establishing a connection. The most common data rates of the voice band modems are 300, 1200, 2400, 4800, 9600, 14400, 19200, 28800 and 33600 bit/s.

The analog signal from a modem is PCM-coded into a 64-kbit/s data stream at the subscriber interface of a local digital telephone exchange. The absolute maximum capacity of the transmission channel through the telephone network can never exceed 64 kbit/s. Some quantizing noise is introduced in the quantizing process of PCM, as we learned in Chapter 3, and it reduces the end-to-end data rate down from the maximum 64 kbit/s. The present highest rate modems operating at approximately 30 kbit/s are quite close to the theoretical maximum when we consider quantizing noise, and we can never develop essentially higher rate voice band modems. The next step is ISDN, which doubles the data rate and makes call establishment essentially faster. Another option is DSL presented in Chapter 4 which makes much higher data rates available.

6.4 Baseband Modems and Leased Lines

6.4.1 Baseband Modems

Baseband modems are not actually modems at all because they do not modulate and demodulate. However, they are called modems because they have a similar purpose in customer premises, that is, they provide data transmission service though the public telecommunications network.

At each end of the baseband circuit we need compatible equipment that encodes and decodes digital data into another digital format that is suitable for cable transmission. This so-called line coding was explained in Section 4.4.

One application of a baseband modem is a corporate network where they are used to interconnect offices in the same area; see Figure 6.9. For this purpose a public network operator leases two cable pairs for a connection between offices. With baseband modems, data transmission rates up to hundreds of kilobits per second are supported. This is often the most economical way to interconnect LANs when the distance is in the order of a few kilometers. LANs are discussed in Section 6.6.

In the case of a long-distance connection it is not economically feasible to build your own dedicated physical connections. This would require repeaters and separate cable pairs or fibers throughout the country. Instead, the required end-to-end transmission capacity is leased from the telecommunications network operator. For long-distance connections the operator uses the same high-capacity optical transmission systems that are used to interconnect public exchanges in the network; see Figure 6.9. The basic data rate unit of the provided transmission rate through the network is 64 kbit/s corresponding to the capacity of one timeslot in the PCM frame; see Section 4.5. This is why the telecommunication carriers provide leased-line services with data rates in multiples of 64 kbit/s.

6.4.2 Leased Lines

The four-wire baseband connection and leased-line long-distance connections that were explained previously are common examples of leased-line connections. The leased line is connected all the time, but dial-up or switched lines are connected only on demand. However, the switched line requires higher investments in the network equipment and the fee is higher if the circuit is connected most of the time. In LAN interconnections the required capacity is often high and the connection is needed so frequently that the leased line usually provides better service with a lower service cost in a regional corporate network.

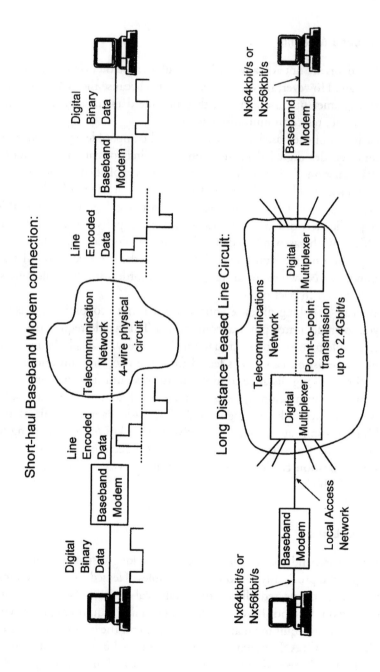

Figure 6.9 Baseband modems and leased lines.

The data rates of baseband modems and leased-line services are higher than that which an ordinary voice band modem transmission can provide. Some examples of the data rates in use are:

56 kbit/s,   64 kbit/s,   128 kbit/s = 2 × 64 kbit/s
192 kbit/s = 3 × 64 kbit/s,   512 kbit/s = 8 × 64 kbit/s

6.5 Integrated Services Digital Network

We introduced the ISDN in Chapter 2 as a new generation telephone network. Now we look at it again from the data transmission point of view. As we noticed in the previous section the leased-line service is not economically feasible if the usage of the service is low. In this case it is often attractive to use a switched telecommunication service to provide the connection only when it is needed. The ISDN provides switched end-to-end digital $n*64$-kbit/s circuits that we may use for voice or data. Figure 6.10 presents an example of an interconnection when ISDN *basic rate interfaces* (BRI), 2*64 kbit/s, are available at both ends of the circuit. Maximum 8 subscriber devices may be connected to NT and two of them may communicate at the same time.

The basic rate interface provides two independent 64-kbit/s circuits, and the routing of one B-channel is independent of the routing of the other channel. This allows residential users to have two independent telephone connections

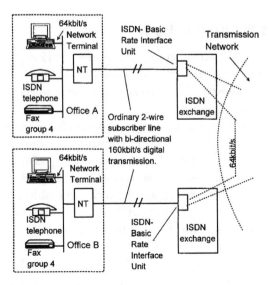

Figure 6.10 A basic rate interface and ISDN connection.

via one two-wire subscriber line or, alternatively, one line for telephone and the other for the simultaneous connection to the Internet. The present fax, group 3 facsimile, operates at a data rate of 9.6 kbit/s. The provision of a 64-kbit/s end-to-end digital connection by ISDN allows quicker and better quality facsimile transmission, and the new standard for this data rate is called group 4 fax. The BRI of ISDN, 2B + D (2 × 64 kbit/s + 16 kbit/s) is aimed to replace the present analog subscriber telephone interface in the future. In corporate networks, higher data rates are required and these are provided by the *primary rate interface* (PRI) that has the structure 30B + D (30 × 64 kbit/s + 64 kbit/s) or 23B + D (23 × 64 kbit/s + 64 kbit/s). The PRI utilizes 2.048- or 1.544-Mbit/s transmission in the local access network and is able to support many simultaneous telephone calls. This interface is used for PABX connections to the public network. The frame structures at 1.544 and 2 Mbit/s were explained in Chapter 4.

6.6 Local Area Network

The data communication systems that we described previously rely on the services provided by a public telecommunications network operator. However, there is a need for high data-rate communications inside a company; and to satisfy these needs local data communication networks, called LANs, are built up.

6.6.1 LAN Technologies and Network Topologies

LANs provide high data-rate communication between computers, for example, inside one building. Because of the high transmission capacity (10 Mbit/s or higher), only short distances are allowed. The typical maximum distance is from a few hundred meters to a few kilometers.

With help of switching devices such as bridges and routers, LANs may be interconnected to make up a wide area corporate network. The bridges interconnect separate LAN segments and route frames from one segment to another with the help of a local hardware address that is stored in the interface unit of each computer. Routers are devices that use network layer addresses for the routing of packets and are used to connect LANs to other networks, for example, to the Internet. Routers can also be used to interconnect LANs that use different technologies.

The basic structures of the two most common LANs, Ethernet and Token Ring, are presented in Figure 6.11. The common principle of all LAN networks is that all computers are connected to the same physical cable and they use it

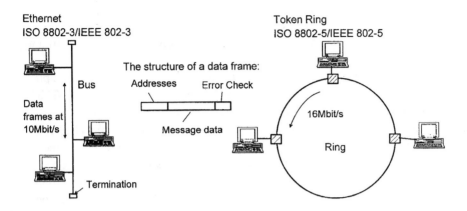

Figure 6.11 Local area network.

in turn. Information is sent in long frames that include the hardware addresses of both the source and the destination. These addresses are unique to each interface card plugged into a computer. Each computer listens to the cable and receives the frames that contain its own identification as a destination address.

Special protocols are standardized to make sure that only one computer transmits at a time. The complex standards of LANs specify the OSI layer 1, the physical layer, and the so-called MAC sublayer of layer 2 (the data link layer). The basic task of these protocols is to connect a computer to another via a shared medium as if they were connected by a point-to-point cable.

The most common LAN is called the "Ethernet" and standardized as ISO 8802-3 or ANSI/IEEE 802-3. Its original principle was developed by Digital, Intel, and Xerox and they called it Ethernet. The standardized protocols are not exactly equal to the original Ethernet, but they can operate in the same LAN.

Another common LAN is Token Ring developed by IBM and standardized as ISO 8802-5 or IEEE 802-5. The typical data rate of this LAN is 16 Mbit/s. In the Token Ring network only a computer holding a special short frame called a token is able to transmit to the ring. The transmitted frame propagates via all computers in the ring and the station with the destination address reads it. The sending computer takes the frame out from the ring and passes the token to the next station in the ring that is then able to transmit. Physically the Token Ring is always built as a star although logically it still makes up a logical ring that was presented in Figure 6.11. All computers are connected to a wire center that bypasses the workstations in a power off condition. When the power is switched on the frames propagate from a workstation via a wire center to the next workstation in a logical ring. The Token

Ring has some technical advantages over the Ethernet (better bandwidth utilization), but it is much more complicated because of the token management and, thus, more expensive.

One important high-speed LAN is the FDDI. Its operation principle is quite similar to the Token Ring, but the data rate is higher, 100 Mbit/s. The FDDI also allows longer distances, and the maximum length of the ring is 100 km. The original transmission media of the FDDI was optical fiber but currently copper cables are also used for connections between computers and a station attachment unit that connects workstations to the ring. The FDDI has been around since the 1980s, and for many years it was the only technology that provided bandwidth higher than 10 or 16 Mbit/s. It was used as a backbone network to interconnect Ethernet or Token Ring LANs. The FDDI technology that supports real-time traffic (FDDI II) such as voice is also available. Now as simpler high-speed technologies are becoming available, the importance of FDDI is beginning to decrease.

There are many other standards for LANs, but the vast majority of LANs in use utilize Ethernet technology because it is simple and inexpensive. In the following subsections we concentrate on Ethernet networks that are presently able to support data rates up to 1 Gbps.

6.6.2 Multiple Access Scheme of the Ethernet

The MAC sublayer of the Ethernet is defined in ISO 8802-3/IEEE 802.3 and this access method is called CSMA/CD. This abbreviation stands for:

- *Carrier sense* (CS) means that a workstation senses the channel and does not transmit if it is not free.

- *Multiple access* (MA) means that many work stations share the same channel.

- *Collision detection* (CD) means that each station is capable of detecting a collision that occurs if more than one station transmits at the same time. In the case of collision the workstation that detects it immediately stops transmitting and transmits a burst of random data to assure that the collision is detected by other stations as well.

The original standard defined thick and thin coaxial cable networks operating at 10 Mbps. Many physical cabling alternatives have been added to the standard, and the twisted pair network 10BaseT is gradually replacing coaxial networks. In response to the increasing need for higher data rates in today's LANs, 100/1000-Mbit/s Ethernet networks are released. The Ethernet offers a seamless path for the development of LANs into higher speeds, while the present infrastructure of the network remains unchanged. In order to support this smooth development of LANs, the latest high-rate networks still use the same frame structure and the same managed object specifications for network management.

We now explain the operation of the CSMA/CD multiple access scheme and the network structure of the original IEEE 802.3. Later in this section we will review the structure and operation of the modern 100-Mbps and 1-Gbps networks.

6.6.3 CSMA/CD Network Structure

For collision detection it is essential to define the maximum delay of the network so that a station can be sure that transmission has been successful or collision has occurred (during transmission). In the case of a coaxial network, each cable segment is terminated by a 50-Ω resistor at both ends to avoid reflections. The maximum length of the cable segments and the number of workstations (or transceivers) connected to each segment are specified. The specifications for coaxial network sections are:

- Thick coaxial cable, 10Base5:
  - Maximum length 500m;
  - Maximum number of work stations 100;
- Thin coaxial cable, 10Base2:
  - Maximum length 200m;
  - Maximum number of work stations 30.

Thick coaxial cable has typically been used in a backbone network that interconnects thin coaxial cable segments into which workstations are connected.

If the network is longer than one cable segment, repeaters may be used to regenerate attenuated signals. Repeaters are physical layer devices that retransmit signals in both directions. Logically the network remains a single physical network where all frames are transmitted to every cable segment; see Figure 6.12.

Collision detection requires that the maximum delay does not exceed a certain value and this restricts how many cable segments can be connected with repeaters. It is defined that the maximum number of repeaters between workstations is four and two of the segments in between have to be link segments that have no workstations.

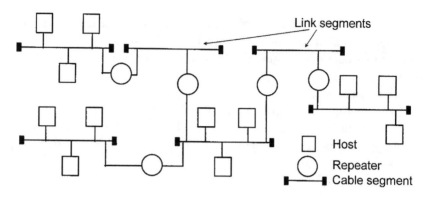

Figure 6.12 Example of coaxial Ethernet network.

If further extension to the network is needed, bridges may be used. The physical size is no more a limitation because physical networks are now isolated from each other by a MAC layer device. It stores and forwards frames according their MAC layer addresses and acts as a workstation interface at each segment.

6.6.4   Frame Structure of the Ethernet

The frame structure of IEEE 802.3/ISO 8802-3 is shown in Figure 6.13. We now explain the purpose and structure of the fields in the frame.

Each frame starts with the *preamble* with 7 bytes each containing the bit pattern 10101010. The Manchester encoding produces a 10-MHz square wave that helps the receivers synchronize with the sender.

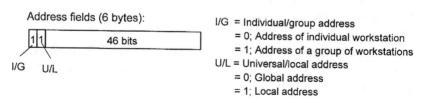

Figure 6.13 Frame structure of the Ethernet.

The *start-of-frame delimiter* contains the bit sequence of 10101011 and indicates the start of the frame.

Both *addresses* contain 6 bytes, with the first bit indicating if it is an address of an individual work station or a group address. Group addresses may be used for multicast where all stations belonging to the same group receive the frame. The second bit indicates if it is a local address (defined locally) or a unique global address. Normally global addresses that are used and they are unique for each network card in any computer. An address range is allocated by the IEEE for each LAN card manufacturer. When a card is manufactured, one of the addresses is programmed into it. This ensures that no two cards will be using the same address in the network. Note that although these addresses are globally unique, they only have local importance. They are never transmitted to other networks. If all stations in a LAN should receive the same message, all destination address bits are set to one. This is called a broadcast address.

The *length-of-data field* tells how many bytes there are in the data field, from 0 to a maximum of 1500. If this number is higher than 1500 in a frame, it cannot be an 802.3 frame. In this case, the frame is a DIX Ethernet frame and a receiver interprets these two bytes as protocol type information that defines a higher layer protocol, for example, the IP.

Data is where the upper sublayer LLC data is carried.

For collision detection the minimum length of the frame is defined to be 64 bytes from the destination address to the checksum. If the data field is very short, the *PAD* field contains random data to extend the frame length to the minimum of 64 bytes.

The *checksum* is used to detect errors when the frame is received. The 32-bit CRC code is used for error detection. If errors are detected, the frame is discarded by the MAC layer and it is left to the upper protocol layers to recover this situation.

6.6.5 CSMA/CD Collision Detection

Suppose that two stations both begin to transmit at the same time to the same cable. The minimum time needed to detect collision is the signal propagation time from one station to the other. However, in a worst-case scenario, the station cannot be sure that it has seized the cable until after two times the propagation delay because the far-end station may transmit just before receiving the signal from the distant station. Then it takes another end-to-end propagation delay until this transmission is detected at the distant transmitting station. As a conclusion, a station can be sure that it has seized the cable and transmitted successfully after two times the worst case propagation delay. As a consequence, to detect collision (before the transmission is finalized) the shortest frame has to be longer than two times the propagation delay. In the case of a 10-Mbps coaxial network the minimum frame length is 64 bytes and correspondingly the maximum length of the network is 2.5 km. The propagation speed is approximately 70% the speed of light.

6.6.5.1 Operation of Collision Detection

The Ethernet transmitter operates as a current generator; see Figure 6.14. When the pulse is transmitted, the current of −82 mA is driven to the cable and the pulse amplitude is −2V (25-Ω impedance). The Manchester line code is used (see Chapter 4) and the average current is about −41 mA when the transmission is on. The average voltage of the cable is monitored by an integrator (low-pass filter) and a comparator that compares the average voltage in the cable with the threshold level, which is set to approximately 1.5V, as shown in Figure 6.14.

If two transmitters are active at the same time, each generates −41 mA on average and (no attenuation is taken into account here) the average voltage is −2V with two active stations at a time. When three stations are active, the average voltage is −3V. If the average voltage goes below −1.5V, the output of the comparator changes state and the collision is detected (multiple stations are transmitting at the same time).

This described principle is specified in the CDMA/CD standard (IEEE 802-3/ISO8802-3). However, actual implementations may perform the collision detection in a different way. They may read signals back from the cable

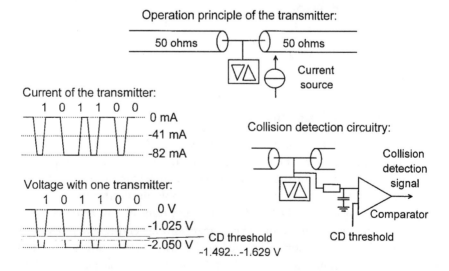

Figure 6.14 Collision detection in Ethernet.

and detect collision if they are different from the originals. They may measure the timing jitter of the pulse edges and detect collision if the edge locations in time do not occur at regular time instants. It is up to the manufacturer of the LAN cards to design the implementation as long as it is equal to or better than that defined in the standard.

6.6.5.2 Contention Algorithm of CDMA/CD

Any station that has a frame to send may transmit at any time if the medium is free or at a transmission instant shown in Figure 6.15. If more than one station decides to transmit simultaneously, there will be a collision. Each station that transmits detects the collision, aborts its transmission, waits for a random period of time, and then tries again (if no other station has started to transmit in the meantime). Therefore, there will be alternating contention and transmission periods, with idle periods occurring when all stations are idle.

6.6.5.3 Binary Exponential Backoff Algorithm

After a collision, time is divided into discrete slots with a length equal to the worst case round-trip propagation time on the network. To accommodate the longest path allowed (2.5 km and four repeaters), the slot time is set to be 512 bit times (51.2 μs) at the data rate of 10 Mbit/s.

After the first collision, each station waits either 0 or 1 slot times before trying again. If two stations collide and each one picks the same random number, they will collide again. After the second collision, each station picks either 0, 1, 2, or 3 at random and waits that number of slot times. If a third collision occurs (probability is now 0.25), then the next time wait time is chosen at random from the interval of 0 to $2^3 - 1$.

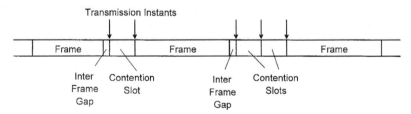

If more than one station transmits simultaneously, contention is detected
and both stations select a random number 0 or 1 and transmit again immediately or
wait for one contention slot (51.2 microseconds).
Depending on how many collisions have occured, random number is selected from
the set 0.....(2^i -1), where *i* is the number of detected collisions.

If 10 or more collisions have occured, selection range is 0...1023.
When 16 collisions have occured, the problem is reported to higher layers.

Figure 6.15 Contention algorithm of CDMA/CD.

In general, after i collisions, a random number between 0 and $2^i - 1$ is chosen and that number of slots is skipped. The probability of the next collision decreases with the number of previous collisions. After 10 collisions have been reached, the randomization interval is frozen at the maximum of 1023 slots. After 16 collisions, the controller reports failure back to the computer. Further recovery is up to the higher layers. The probability of this situation is so small that it does not occur in normal operation, but it may happen for example if the coaxial cable is cut off. Then each transmitted frame is reflected from the broken end of the cable and collision is detected for each transmission.

This described algorithm, called binary exponential backoff [3], was chosen to dynamically adapt to the number of stations trying to send. If a number of stations trying to send is high, a significant delay will result. However, if the stations would only have options 0 or 1 to choose, and if there were 100 workstations, it would take years to have a successful transmission.

There is no simple mathematical solution to estimate delays of the CDMA/CD accurately. Practical experience has proved that to have reasonable performance out of 802.3 the loading has to be kept in the order of 30% or less of the maximum physical data rate on average.

The CSMA/CD as a MAC sublayer operation provides no acknowledgments that garbled frames have just been discarded. If acknowledgments are used by higher protocol layers, they appear just like other frames in the network.

Figure 6.16 shows an example where there are three active stations in the CDMA/CD network. At time instant 0 both stations A and B transmit simultaneously and collision is detected. Then station A decides to transmit again but station B decides to wait for one slot time. Station C transmits at the same time as station A and a second collision occurs.

Now both A and C decide to skip one slot and station B seizes the network. Both A and C transmit when the network is free again and a third collision occurs. Now station A has suffered from three collisions and its range for the second transmission is 0 to 7 slots. Station A now has a wide range and selects most probably a higher number than station C that has had only two collisions.

6.6.6 Twisted Pair Ethernet

The twisted pair CDMA/CD network called 10Base-T uses a twisted pair to connect workstations to the wire concentrator. Twisted pair is easier and more flexible to install than coaxial cable, and this has made 10Base-T very popular. In the most simple structure the concentrator or hub acts as a repeater transmitting frames from one workstation to all the others, as shown in Figure 6.17.

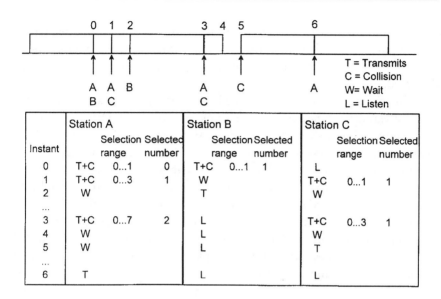

Figure 6.16 Contention example.

Instant	Station A			Station B			Station C		
	Selection range	Selected number		Selection range	Selected number		Selection range	Selected number	
0	T+C	0...1	0	T+C	0...1	1	L		
1	T+C	0...3	1	W			T+C	0...1	1
2	W			T			W		
...									
3	T+C	0...7	2	L			T+C	0...3	1
4	W			L			W		
5	W			L			T		
...									
6	T			L			L		

Legend: T = Transmits, C = Collision, W = Wait, L = Listen

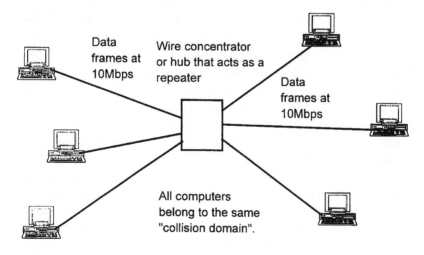

Figure 6.17 Twisted pair shared media CDMA/CD.

In 10Base-T network collision detection operates exactly the same way as we described previously as long as we ensure that the worst case propagation delay from one station to another does not exceed half of the transmission time of the shortest frame. The twisted pair restricts the transmission distance between a concentrator and a workstation to approximately 100m. At 10 Mbit/s the maximum distance between workstations is the same as in the coaxial networks and networks may consist of multiple interconnected writing concentrators. Note that when the data rate is increased, the duration of the shortest frame is decreased and the maximum distance is correspondingly decreased.

In the network that we described previously all frames are transmitted to every segment in the LAN. We call this *shared media* CDMA/CD because all computers use a share of the transmission capacity of all segments. The shared media networks shown in Figure 6.12 and 6.17 make up a single "collision domain," that is, collision occurs if two or more computers anywhere in the network transmit so that two or more frames overlap. Bandwidth utilization of the shared networks is poor because one transmitting workstation seizes the whole network although only the segments to the source and destination computers are needed for communication. The switched LAN, which is discussed next, divides a LAN into multiple collision domains and the bandwidth utilization is much improved.

6.6.7 Switched Ethernet, Switches, and Bridges

We can improve the performance of a CDMA/CD network using switches instead of repeaters. Switches have become popular in all CDMA/CD networks including coaxial networks where we call them bridges. Bridges do not transmit all received frames to all ports like repeaters. Instead they use MAC addresses and transmit frames to the direction where the destination is known to be located. Bridges are able to learn by listening to the traffic. They read the source address in each frame and build up a routing table containing all stations that have transmitted a frame. If the location is not yet known, frames are transmitted to all ports. The routing table is updated continuously to allow a workstation to move from one port to another.

In the network topology shown in Figure 6.17 we may change the repeater to a switch that is actually a fast multiport bridge. Now the frames from one computer to another are transmitted only from the source port to the destination port and two other computers connected to different ports may transmit to each other at the same time. To allow this the internal capacity of a switch is much higher than the data rate at one port.

We may connect many repeater hubs shown in Figure 6.17 to each other by a switch or a switching hub. The switches create separate collision domains since they do not forward collision signals from one port to another.

6.6.8 Fast Ethernet

The most important Fast Ethernet standard is 100Base-T that carries data frames at 100 Mbps. The data rate is increased by a factor of ten but the frame format and MAC mechanism remain the same as in coaxial Ethernet and 10Base-T. The fast Ethernet specifications include a mechanism for the Auto-Negotiation of the medium speed, making it possible for vendors to provide dual-speed Ethernet interfaces that can be installed and run at either 10 Mbps or 100 Mbps automatically.

The topology of the 100Base-T network is equal to the 10Base-T shown in Figure 6.17. Connections between workstations and a repeater are twisted pairs. The standards include both full-duplex connections where one pair is used for each transmission direction and half-duplex connections where a single pair is used in both directions. Both shielded and unshielded twisted pairs are specified as transmission media. The segment length is limited to a maximum of 100m to ensure that round-trip timing specifications are met. The fast Ethernet also specifies optical fiber connections that allow longer distances than a twisted pair.

Just as in the case of the 10Base-T, we can improve the performance of the 100Base-T network using switches instead of repeaters (or Hubs). With the help of switches that support both data rates we can gradually update the network and increase the data rate only where it is required. As a next step, the Gigabit Ethernet, which we describe next, further increases the capacity of the Ethernet networks. It provides a smooth path to gradually increase the performance of Ethernet LANs.

6.6.9 Gigabit Ethernet

The Gigabit Ethernet provides 1-Gbps bandwidth with the simplicity of Ethernet at a lower cost than other technologies of comparable speeds. It will offer a natural upgrade path for current Ethernet installations, leveraging existing workstations, management tools, and training.

Gigabit Ethernet employs the same CSMA/CD protocol, frame format, and size as its predecessors. Since the Ethernet is the dominant technology for LANs, this means for the vast majority of users that they can extend their network to gigabit speeds at a reasonable initial cost. They need not re-educate their staff and users and they need not to invest in additional protocol stacks.

The Gigabit Ethernet in an efficient technology for backbone networks of Ethernet LANs because of the similarity of the technologies. As an example, for an ATM backbone network the frames of the Ethernet must be translated into short ATM cells and vice versa. The Gigabit Ethernet backbone transmits Ethernet frames just as they are. The only difference is the data rate.

The Gigabit Ethernet may operate in full-duplex mode; that is, two nodes connected via a switch can simultaneously receive and transmit data at

1 Gbit/s. In half-duplex mode it uses the same CSMA/CD access method principle as the lower rate networks.

6.6.9.1 Extension to Collision Detection Method

The Gigabit Ethernet CSMA/CD method has been enhanced in order to maintain a 200-m collision diameter at gigabit speeds. Without this enhancement, maximum-sized Ethernet frames could complete transmission before the transmitting station senses the collision, thereby violating the CSMA/CD method. Note that the duration of a frame is now only 1% of that at 10-Mbps data rate.

To resolve this issue, both minimum CDMA/CD carrier time and the Ethernet slot time have been extended from 64 to 512 bytes. The minimum frame length, 64 bytes, is not affected; but frames shorter than 512 bytes have an extra carrier extension. This so-called packet bursting affects small-packet performance but allows servers, switches, and other devices to send bursts of small packets or frames to utilize available bandwidth. Devices that operate in full-duplex mode are not subject to the carrier extension, slot time extension, or packet bursting changes because there are no collisions.

6.6.9.2 Applications of Gigabit Ethernet

As we have seen, the Gigabit Ethernet offers a smooth transition of a LAN to higher bit rates where they are needed. Often a greater bandwidth is first needed between routers, switches, and servers. Figure 6.18 shows an example of how different Ethernet technologies can be used in the same network.

In the example we originally had a 10Base-T network where the workstations were connected to repeaters. As capacity demand increased, repeaters were upgraded to switches at the same data rate or repeaters or switches at 100 Mbit/s. If 10 Mbit/s is high enough for an individual workstation, there is no need to update the network interface card of that computer as long as the port of the switch supports that data rate. The two 10/100-Mbit/s switches are 10-Mbit/s switches that were upgraded with a network interface card that connects them at 100 Mbit/s to the higher level switches in the network. The highest level switches were upgraded with 1000-Mbit/s cards for their interconnections and for the connections to the servers in the same way. In the next upgrade we would probably replace highest level switches with genuine gigabit Ethernet switches and use the old ones to replace lower level switches or repeaters.

There are many possible transmission media for the gigabit Ethernet such as multimode optical fiber, 1000Base-SX/LX, single-mode optical fiber 1000Base-LX, coaxial cable 1000Base-CX, and category 5 pair cable 1000Base-T.

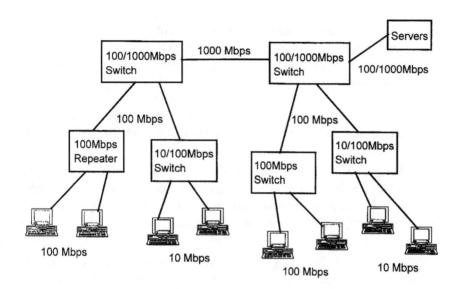

Figure 6.18 Ethernet network operating at 10, 100, and 1000 Mbit/s.

6.6.10 Virtual LAN

A large physical LAN can be divided into many logical LANs. This improves the performance and security of the LAN because the traffic of, for example, a marketing department is separated into a dedicated logical LAN. A straightforward way to define a *virtual LAN* (VLAN) is that all computers connected to certain ports of a switch make up one logical LAN and traffic is switched between these ports only. Another more flexible way to configure a VLAN is to define the MAC addresses of all computers that belong to a certain network. This principle is more difficult to manage because a network manager has to define a VLAN for each MAC address. On the other hand, switches may dynamically configure themselves when a computer is transferred to a new physical location. This principle also allows a computer to belong to many VLANs. The third way to configure VLAN is define all computers using a certain network layer protocol to make up their own VLAN. The VLAN of a certain protocol may be further divided into smaller VLANs by defining a certain set of the network layer addresses that make up a VLAN.

6.7 Circuit and Packet Switching

In the preceding sections we described many principles for data communication through the telecommunications network. We may divide these data connections into different categories based on the principle of how the communication circuit is built up between the communicating devices.

Data and voice communication through the telecommunications network may use two basic different types of circuits:

- *Leased* or *dedicated*: the cost of which is fixed per month and depends on the capacity and length of the connection. Leased lines were described in Chapter 6.4.

- *Switched* or *dial-up*: the cost of which depends on the time the service is used or the amount of transmitted data and on the distance. There may be many other parameters that influence the cost such as the maximum data rate or average data rate.

Within the switched category there are two subcategories: circuit- and packet-switched networks (see Figure 6.19)—both are used for data transmission. Figure 6.19 shows what switching principles are used by the networks that we have discussed previously, and will discuss later in this chapter.

6.7.1 Circuit Switching

Circuit-switched networks provide fixed bandwidth and very short delay. It is the primary technology for voice telephone, video telephone, and video conferencing. The disadvantage is that it is inflexible for data communication

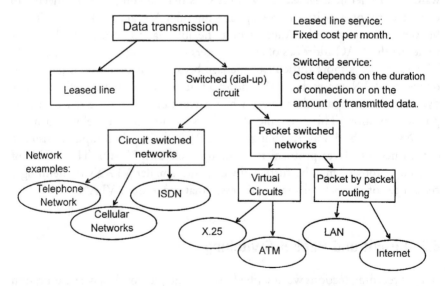

Figure 6.19 Leased lines and circuit and packet switched networks.

where the demand for the transmission data rate is far from constant but varies extensively over short time scales.

Some older generation data networks use the circuit switching principle. In the beginning, a circuit-switched connection is dialed-up by the data source. The routing is based on the destination subscriber number. The connection is released after the communication is over; see Figure 6.20. During a conversation the data capacity of the connection is fixed and it is reserved only for this conversation regardless of whether the data capacity is used or not. At the end of the call, the circuit is released. ISDN as well as the telephone network use the circuit switching principle.

6.7.2 Packet Switching

Packet-switched networks are especially designed for data communication. The source data is split into packets containing route or destination identification. The packets are routed toward the destination by packet switching nodes on the path through the network. The major drawback of the packet-switched technology is that it usually cannot provide a service that requires constant and low delay.

There are two basic principles of packet-switched networks as illustrated in Figure 6.20, namely, virtual circuits and datagram transmission.

In the case of the virtual circuit, the connection is established in the beginning of each conversation and every packet belonging to a certain connection is transmitted via the same established route. This route is called the *virtual circuit*. The PSPDN X.25 belongs to this category as well as ATM and Frame Relay networks. The difference between circuit-switched physical circuits and virtual circuits is that the capacity of the transmission lines and channels between network nodes is shared by many users. At a certain moment active users may use all the capacity if other users do not transmit anything. The complete address information is not needed in the packets after the connection has been established. Only a short connection identifier is included in each packet to define the virtual circuit to which the packet belongs. The operation of switched virtual circuits is explained in more detail in Section 6.7.4.

Another method for packet-switched data communication is connectionless datagram transmission where routing devices, perform routing procedures, and each packet contains a full destination address. We discuss this layer 3 routing principle next.

6.7.3 Layer 3 Routing and Routers

In the case of layer 3 routing every data packet carries complete global destination information (network layer address of the destination) and all packets are

Circuit Switched Data Transfer

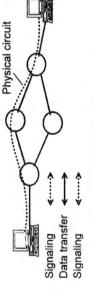

The circuit is first established, then data is transferred and in the end the circuit is released. The capacity of the circuit is not available for other users.

Examples: telephone network, ISDN

Packet Switched Data Transfer

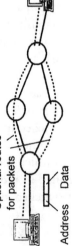

In true packet switched data communication there is no dedicated connection between communicating devices. Each packet includes complete destination address and is sent and routed independently.

One example is the Internet.

Packet Switched Data Transfer with virtual circuits

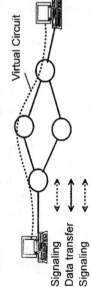

Virtual circuit is established in the beginning, then data is transferred via the same path and in the end the virtual circuit is released. Each packet include circuit identification. Capacity between nodes is shared by all users.

Examples: X.25, Frame Relay, ATM

Figure 6.20 Circuit and packet switched data transfer.

routed independently. As a consequence, each packet may use a different route and arrive out of sequence. The operation principle of the Internet belongs to this category.

The routing procedure is performed at the network layer, layer 3, and requires analysis of each packet and the routing decision based on the destination address. Packets are stored and, when the route and the corresponding port of the router are defined, forwarded to the next router on the path to the destination. This principle of operation makes routers slower than the switching devices that operate at data the link layer, layer 2. The operation principle of the data link layer switches is discussed next.

6.7.4 Switching and Routing Through Virtual Circuits

The routing of packets is based on the virtual circuits in most public data networks such as X.25, Frame Relay, or ATM networks. Each frame or cell on a virtual circuit contains identifying information about the circuit to which it belongs. This identification has a different name in each network, but we call this identification the *virtual circuit identifier* (VCI). During the circuit establishment phase, signaling messages are exchanged between user equipment and a network. In the network each circuit established between nodes has a certain identification number and there is no global identification that could be used on all the links through the network. Instead, one of the free circuit identifications on each link is allocated for a certain virtual circuit under establishment and the routing tables of switching nodes in the network are updated to contain all established circuits; see Figure 6.21.

VCIs only have local significance on a specific network link and, therefore, are changed as a frame traverses the virtual path through the network.

When a frame is received from a certain link, the frame switch simply reads the VCI and combines the incoming link number to determine the corresponding outgoing link and VCI. The new VCI is then written into the frame header and the frame is queued for forwarding on the appropriate link. The order of frames is preserved and routing them is very fast because it does not require analysis of a global address.

In Figure 6.21 a frame switch has routing tables for each incoming data link. Let us assume that a frame with identification 3 is received from the link 3. The switch looks up the link 3 routing table and finds out that this frame should be transmitted to link 2, so identification 3 is replaced by 1.

This process is fast because it does not require any network layer routing with a global address. Instead this routing is done in the data link layer. The VCI is also very short and the utilization of data capacity in this kind of

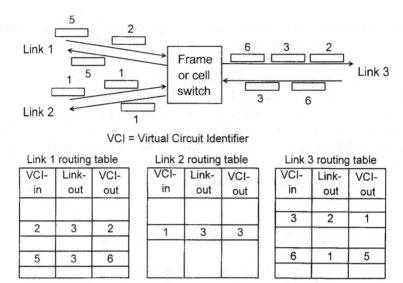

Figure 6.21 Routing of frames on virtual circuits.

network is more efficient than if the global address was included in each frame or packet.

6.7.5 Gigabit Switching Routers

As we have seen, routers are store and forward devices that operate at the network layer and use the global destination address of the packet for routing. The data link layer switches use local identifiers and are able to switch packets at wire-speed. The switched LANs use the data link layer MAC addresses, and the Frame Relay and ATM networks use the VCIs for switching, which requires that the tables containing VCIs (as shown in Figure 6.21) or MAC addresses are available for the data link layer; the network layer routing function need not be involved.

High-speed backbone networks utilize the fast Ethernet, gigabit Ethernet, FDDI, Frame Relay, or ATM technology. On the other hand, the IP, which has been transmitted through the traditional router networks, carries and will carry most of the user data. Clearly for IP packets a more efficient method than layer 3 routing is required to utilize the full capacity of the high-speed backbone networks and to speed up the growing Internet traffic.

There are many evolving technologies, such as Multiprotocol over ATM (MPOA of the ATM Forum), Fast IP of 3Com, Tag Switching of Cisco, gigabit routers, multigigabit routers, cell switch routers, and IP switching addressing the high speed switching of the network layer packets, especially

IP packets. We don't discuss these technologies here in detail but just briefly introduce the main two principles that they use to speed up routing.

The network layer that looks outward at the topology and hierarchy of the network calculates the routing tables, which are then used for routing to avoid the need to send every packet through a centralized routing processor of the router. The routing tables that contain the network layer addresses for high-speed packet-by-packet routing are given to the interface units of the router. Now they carry out routing at each port in parallel to improve the performance.

In many of the aforementioned new technologies the network layer routing is only performed for short-term packet flows, and for long-term flows the table containing the data link layer addresses or circuit identifiers is created. These tables are used for high-speed switching at the data link layer, as explained in Section 6.7.4. This principle speeds up routing to wire-speeds and can be used with all backbone network technologies. However, many vendors are focusing on ATM as a backbone technology and use cell switching in their high speed routers. This ATM switching of IP traffic is called "IP switching" and is introduced in the last section of this chapter.

6.8 Internet and TCP/IP

The Internet has developed into the world's major information network and this will continue. Here, we review its development and the most important protocols on which the operation is based.

6.8.1 Development of the Internet

The worldwide Internet network developed from experimental computer networks in the 1960s to the worldwide university network in the 1970s and 1980s. The original technical development was supported by the *Department of Defense* (DoD) of the United States. The Internet technology is not as formally standardized as other public telecommunications networks. There is no standardization body such as the ITU-T where all nations together participate in the development of the network. However, some centralized control is required and there is an organization that manages the development of the Internet. This institution is the *Internet Engineering Task Force* (IETF) in the United States and it updates standards, informs about changes, and controls the usage of global addresses of the Internet. Technical specifications of the Internet are called *Requests For Comments* (RFCs) instead of standards. This freedom in development of the Internet has accelerated the growth of the network.

The Internet has been used by the academic society for twenty years. It used to be difficult to use, only some organizations had access to it, and the only users were academic specialists that were familiar with it. Since there were no commercial applications, the *usage charges* were not considered at all in the Internet technology. The academic information exchanged over the Internet was public by nature and security functions were not considered in the development of the Internet.

The development of a graphical user interface exploded the usage of the Internet in the middle of 1990s. This new user interface is called the WWW and it has made the Internet easy to use for anyone. There are currently many commercial *Internet service providers* (ISPs) who have access to the worldwide Internet service available for ordinary telephone or ISDN subscribers. Anyone who has a personal computer may access the Internet via the telephone or ISDN network.

The Internet was originally designed for data applications only and it uses the packet-switched transmission principle that we explained in the previous section. This is a very efficient method because the transmission connections in the network are used on demand. There is no circuit and fixed share of capacity for each user as we have, for example, in ISDN. Because of this efficiency, it is expected that the Internet will be used more and more instead of PSTN for voice communication. The usage of the Internet for international calls is very attractive because they are often free of charge. There is no method for charging a certain type of usage of the Internet and the user's fee is typically based on the time they are connected to the network of their ISP. However, because of the variable delay of packets, the quality of speech is presently not as good as in PSTN.

The Internet has developed to the major information network in the world. There are still some problems with its technology that restrict its usage. The major problems are: charging for services, security, quality of the interactive real time information, capacity of the network when usage increases, and a shortage of the Internet addresses. However, the technical solutions for these problems are under development or implementation, and the rapid growth of the Internet is expected to continue with more and more commercial applications becoming available. We may also expect that the Internet will take a share of the telecommunications for which we presently use PSTN or ISDN.

6.8.2 Internet Protocol

The main task of the IP is addressing, which requires global Internet addresses and routing of the IP packets from the source computer to the destination. The basic network elements in the network are routers in the network and permanently connected computers (hosts) with different application protocols that provide services. Each such element has at least one Internet address that is different from the addresses used in the PSTN. The Internet addresses are global and their usage is internationally controlled by the IETF. Figure 6.22 shows the principle of the hierarchical routing between computers that are connected to different networks such as separate LANs. As we see the routers may have multiple IP addresses; in the figure there are three subnetworks that makes packet routing hierarchical.

An IP address consists of four numbers between 1 and 255, for example 130.237.21.11. Applications have their own more user-friendly way of addressing. A widely used application example is e-mail; for example, we may have an e-mail address such as tarmo.anttalainen@evitech.fi where the last section, the domain name, is translated in the network to the IP address number. The Internet routes e-mail messages according to that address number to the computer where the mailbox of the receiver is located.

The IP provides global routing that is based on IP addresses. The communication over LANs is based on the local addresses. When a computer in a LAN is connected to the Internet, it uses these local addresses inside a LAN in addition to the IP addresses. The IP contains methods to determine which local address corresponds to a certain IP address in a LAN. This local address is attached to a data frame that is then transmitted through the LAN. The receiving router of that LAN, via which the Internet access is provided, discards the local address and uses the IP address to further route the message through the Internet.

6.8.3 Transmission Control Protocol

The most important protocol next to the IP is the *Transmission Control Protocol* (TCP). IP provides connectionless transmission through the network, meaning

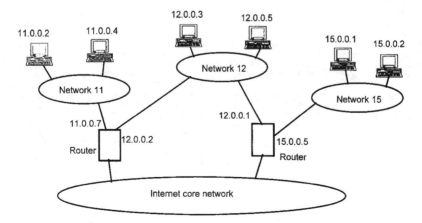

Figure 6.22 IP addressing on the Internet.

that it routes each packet of data independently using the IP address in each packet. Most applications require that the packets arrive in the original order; if one of them is in error, it must be transmitted again. The procedures that are required for these functions are implemented in the TCP protocol. The TCP that runs only in the data source and destination machines provides actually connection-oriented reliable communication over the Internet. In order to do this it establishes a logical connection, checks out if errors have occurred in packets, retransmits packets in error, and rearranges the packets if they arrive out of order.

6.8.4 Application Protocols

The TCP/IP provides reliable data transfer over the Internet. The application protocols are specific for each service. The most important of these are Internet e-mail, *File Transfer Protocol* (FTP) and *HyperText Transport Protocol* (HTTP) for the WWW. Figure 6.23 shows how these protocols are used in the internetworking between two hosts connected to different LANs.

As we see in Figure 6.23, the two intermediate routing devices contain only protocols up to the IP layer. That is enough because their tasks are to route the IP packets from the source to the destination. The end systems contain higher layer protocols as well. The TCP is used by the application protocols and provides a reliable connection between end systems through LANs and the WAN. When we are using, for example, a WWW service, with the help of the TCP we feel that we have fixed reliable connection in use although some IP packets may be lost on the way. The TCP recovers lost packets by retransmitting them. Application protocols run on top of the TCP and typically are different for each purpose, for example, FTP for file transfer and HTTP for WWW.

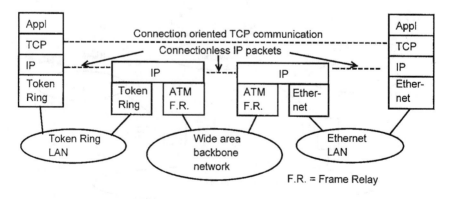

Figure 6.23 Example of Internet networking.

In the example of Figure 6.23 there are many different networks in use and the computers supplied with the TCP/IP protocol can use different LAN technologies. For example, in the backbone network for example ATM or Frame Relay technologies can be used for transmission of IP packets.

6.9 Fast Packet Switching and ATM

Fast packet switching usually means a switched transmission of data frames or packets at 1.5 Mbit/s, 2 Mbit/s, or higher data rates. The conventional global packet-switched service X.25 has a maximum data rate of 64 kbit/s that is rather limited for the applications of today. New services for high rate data transmission have become available. The two most important of these are Frame Relay and ATM.

6.9.1 Frame Relay

Frame Relay transmits data frames with variable length just as X.25 does. The X.25 packet-switched network was originally designed for a low-quality physical network and includes data integrity checking at many protocol layers. With the present high-quality physical network it is usually unnecessary. Frame Relay leaves data checking and acknowledgment procedures to network users and the protocols in use are much simpler and can support a much higher data rate up to 50 Mbit/s.

Frame Relay technology is used to provide semipermanent connections for LAN interconnections. The protocol is based on ISDN signaling protocol LAPD and is defined in the ITU-T/CCITT recommendations I.122/Q.922. Switched Frame Relay service is currently provided by some networks of today. The recommendations Q.931/I.450/1 define the sequence of messages that are exchanged over the D-channel for connection establishment.

Frame Relay is a technology for data transmission and does not support isochronous transmission, such as voice or video, that requires low and constant delay. A new and fast network technology that is designed to support isochronous services as well is known as ATM.

6.9.2 Asynchronous Transfer Mode

All current packet-switched techniques (LANs, X.25, Frame Relay) make use of variable-sized packets, and this leads to significant variations in the arrival times of the packets of a particular data stream.

Since each physical connection may carry traffic from many individual data streams, it occurs every now and then that a specific packet is queued behind a number of large packets from other data streams that are waiting to be sent out on the physical connection. A further consequence is that switching is carried out by software that will eventually constrain the speed and performance of the network.

ATM is one of the cell relay technologies that use small fixed-sized frames called cells. Cell relay transmits frames with constant length, 53 octets, and provides both *variable bit rate* (VBR) *service,* which is optimum for data transmission, and *constant bit rate* (CBR) *service* for voice and video applications. CBR is not available in Frame Relay or X.25.

ATM defines the structure of cells, continuous transfer of cells, and cell switching. Isochronous service is available by reserving a certain fixed capacity of ATM cells from the network. ATM is becoming popular as a high-rate data communication technology. Later it will be used as a transmission method in *broadband-ISDN* (B-ISDN) that will provide high subscriber data rates. ATM cells are packed into an SDH-frame, STM-1, or into SONET frame and then the physical data rate may be 155 Mbit/s or higher.

Significant advantages of the cell relay technology follow from the use of fixed-sized small packets or cells instead of packets with variable lengths. The consequences of this principle are:

- Delay in the network is much lower and more predictable. By ensuring that the cells from a specific data stream occur at regular intervals in the cell stream, it is possible to provide guaranteed bandwidth with low delay and jitter just as in circuit-switched networks.

- The fixed size of cells allows the switching function to be removed from software into hardware with a dramatic increase in switching speed. The current X.25 network supports data rates up to 64 kbit/s; Frame Relay supports up to some megabits per second. ATM is aimed to support data rates from some hundreds of Megabits per second to Gigabits per second.

ATM thus provides the benefits of circuit- and packet-switched networks, allowing all types of traffic to be integrated onto a single network. This is why ATM is defined to be used in B-ISDN in the future. The first implementations of ATM are high-rate data applications. In many present ATM networks the switches are configured to provide semipermanent data connections, by which we mean that these connections are dialed up by users but controlled from the network management center by a network operator.

6.9.3 Protocol Layers of ATM

ATM networks can be considered as a number of layers providing different functions. The three layers that make up an ATM stack are physical layer, ATM cell layer, and ATM adaptation layer, as shown in Figure 6.24.

Standards for the physical and cell layers are presently defined, but the adaptation layer that provide options depending on application requirements is still under development.

ATM networks are connection-oriented, which means that a call establishment phase is followed by a data transfer phase. During an establishment phase a path (virtual circuit) through the network is built up and this path is then used by all cells of this call. ATM thus provides a guarantee of cell sequencing, but some cells in a data sequence may be lost. The cells with errors are discarded by the network, and it is up to the end systems to detect and recover from a cell loss. The control of virtual paths and circuits is carried out by signaling on the subscriber interface called a *user network interface* (UNI). The interfaces between nodes in the network are called *network node interfaces* (NNI).

6.9.4 Cell Structure of ATM

The development of ATM standards is ongoing in such organizations as ATM-forum, ITU-T, ANSI, and ISO. The ATM-forum is working for the fast development of these networks and produces guidelines for implementation to ensure interoperability. All major manufacturers of ATM-systems participate in this work.

ATM = Asynchronous Transfer Mode

B-ISDN protocol reference model:

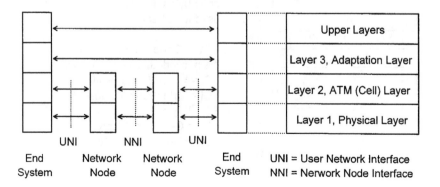

Figure 6.24 The protocol layers of ATM.

The ATM cell is 53-bytes long with 48 bytes to carry the payload and 5 bytes for the header; see Figure 6.25. This size was a compromise between 64 bytes favored by data community and 32 bytes preferred by the voice community. The data fields of a cell are shown in Figure 6.25 and explained in the following subsections.

6.9.4.1 Generic Flow Control

The *generic flow control* (GFC) field is used at a user interface only to control the data flow between the first ATM switch and the user node. Inside the network, that is, in the NNI, this field is used for virtual path identification together with the other VPI fields. This is the only difference in the cell structure of UNI and NNI.

6.9.4.2 Virtual Path Identifier and Virtual Channel Identifier

The majority of the header is taken up by VPI and VCI. Together they identify an individual connection. They have only local significance and change as they proceed through the network, as explained in Section 6.8. They are used the same way as the *logical channel number* (LCN) of X.25 or *data link connection identifier* (DLCI) of Frame Relay.

6.9.4.3 Payload Type

Payload type (PT) specifies whether the cell contains user information or information to be used by the network itself, such as O&M. The network can use these maintenance cells between nodes to perform operations, for example, in a congestion situation.

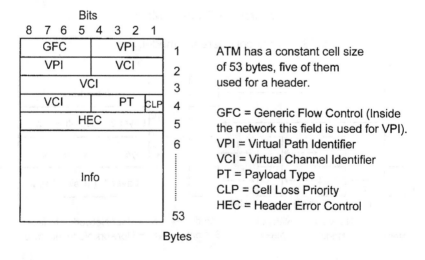

Figure 6.25 The structure of an ATM cell.

6.9.4.4 Cell Loss Priority

The *cell loss priority* (CLP) field carries information between an ATM user system and the network. For example, in a congestion situation the network may use this field to define the priority of cells in the queues or to decide which cells are discarded first in the case of overload.

6.9.4.5 Header Error Control

Header error control (HEC) is a checksum for the first four bytes. It makes possible the detection of multiple errors and the correction of a single error. ATM cells with more than one error will be discarded by the network. It is up to the end systems to detect and recover from such losses. The end systems also have to detect errors in the user data. When an ATM switch updates the virtual circuit and path identifications of a cell, it calculates a new HEC.

6.9.5 Physical Layer of ATM

ATM cells can potentially be carried over most physical layer media, but the ITU-T has defined SDH to carry ATM cells with speeds of 155.52 and 622.08 Mbit/s. Figure 6.26 illustrates how ATM cells are inserted into the payload of an SDH frame STM-1. The frame includes 9 × 290 bytes and is transmitted 8000 times a second. We review here only the STM-1 frame of

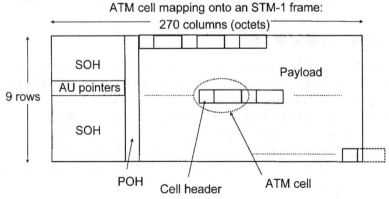

SDH interfaces STM-1 and STM-4 are defined to carry ATM cells.
Many other interfaces are also defined and under study.

ATM cell mapping onto an STM-1 frame:
270 columns (octets)

Continuous stream of cells; dummy cells are inserted if there is no traffic.
Cell synchronization is based on an error check code in the header.

Figure 6.26 Physical layer of ATM.

SDH as an example; this principle is valid for SONET as well, but the detailed structure of the frame is different.

The SDH frame includes *section overhead* (SOH) that contains framing information, network management data, and other overhead information needed by optical SDH transmission systems. ATM cells are merged together with *path overhead* (POH) to the payload of an SDH frame. POH and data (ATM cells) together make up the *virtual container* (VC) that may start at any point inside the payload of the frame. The *administrative unit* (AU) pointer tells where the frame (VC) containing POH and ATM cells starts inside the payload [4]. In Figure 6.26 this starting point is assumed to be in the beginning of the payload.

At an SDH interface the concept of continuous stream is used and dummy cells that are called idle cells are inserted if there is no traffic.

In addition to SDH the ATM cells can be carried in many other transmission systems such as DS-1 (1.544 Mbit/s), E-1 (2.048 Mbit/s), or SONET.

6.9.5.1 Cell Synchronization

Since the cell stream is continuous, there are no explicit indicators for the start and end of a cell. The synchronizing scheme relies on the *check code* (HEC) in the frame header. The receiving equipment calculates the code as each bit arrives and checks if the next byte corresponds to the calculated result. If they are the same it is likely that a cell header has just been received. If the codes of a few cells match, the synchronization is accepted and the equipment (ATM system) becomes operational.

If more than a defined number of cells arrive with a wrong code, the equipment assumes that synchronization is lost and attempts to re-establish it.

6.9.6 Switching of ATM Cells

When an ATM call is established through the network, the cells of this call carry a certain virtual path identification that is changed for each link between switching nodes, as explained in Section 6.7. The circuit identification is divided into two parts and ATM switches can operate at two levels: virtual path level (a group of virtual channels) and/or virtual channel level. Virtual paths act as pipes for a collection of VCs. ATM switches may act at the VP level, the VC level, or both. A VP switch does not look at the VCs within a path and end systems can freely establish and remove VCs without the network carrying VPs being involved.

The routing at the VP level allows the network operator to provide a virtual private network for a corporation just by crossconnecting virtual paths between offices. These paths provide the permanent connections between

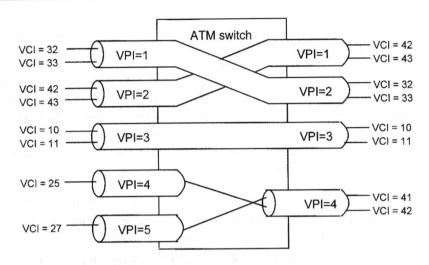

Figure 6.27 Virtual paths and channels of ATM.

offices. The user may then configure his private network the virtual channels inside the virtual paths he has leased from a network operator.

To establish an individual call through a switched ATM network the signaling cells with a specific VCI (reserved for signaling purposes at a UNI) are exchanged between the network and the user. Then the network defines a route with a certain VPI and VCI for this call in each link. To route cells the ATM switches look at both the VPI and VCI in each cell.

In Figure 6.27 virtual paths 1, 2, and 3 are crossconnected at path level and virtual channel identifications remain unchanged. The figure also shows an example where virtual channels 41 and 42 of virtual path 4 are switched to the virtual channels 25 and 27 of virtual paths 4 and 5. In this case the switch updates both the VPI and VCI fields in the header of ATM cells before it transmits them to the next switch.

6.9.7 Service Classes and Adaptation Layer

ATM is designed to support different services and the purpose of the *adaptation layer* (AAL) is to make the cell transport of ATM suitable for different applications. The service classes are defined to support various types of applications. The corresponding AAL protocols define how each class of service is implemented. The role of the AALs is to provide the mapping of particular types of traffic onto the underlying ATM cell layer. This requires that the AAL header containing the protocol information be added to the user data before transmission in the payload of ATM cells shown in Figure 6.25. Depending on the AAL in use, this reduces the user information from 47 to 44 octets in a cell [5]. We can divide services into four basic classes depending on whether the required service is a constant or variable bit rate; isochronous or synchronous; or connection-oriented or connectionless.

Figure 6.28 shows the service classes, their basic characteristics, and corresponding AAL protocol. The timing relation characteristic tells if the information about timing has to be available for the receiver of ATM cells. This may be required for example in the case of the reconstruction of PCM-coded speech, which requires that samples arrive at regular intervals. There are four service classes supporting different applications:

- Class A: Constant bit rate service for voice and video application;
- Class B: Variable bit rate service with timing information for variable bit rate voice and video application;
- Class C: Variable bit rate service for ordinary data application;
- Class D: Variable bit rate service for connectionless transmission of short data messages.

6.9.7.1   Class A/AAL1

AAL1 provides the support for traffic, which requires *constant bit rate* (CBR) service mainly for voice and video applications. This AAL is quite simple because there is no requirement for error detection and recovery for this type of traffic. The transfer of timing information over a call is the major function that AAL1 has to provide to ensure the synchronization of data. This service simulates leased-line data or voice circuits, and one important application will be PABX voice channels in integrated corporate networks that utilize ATM technology.

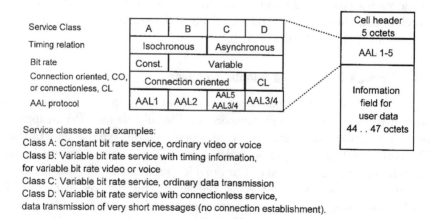

Figure 6.28 Service classes and adaptation layer of ATM.

6.9.7.2 Class B/AAL2

AAL2 provides the support for variable bit rate traffic that requires maintenance of timing information during the call. Timing information is transmitted in the AAL header. Examples of this type of traffic are variable bit rate voice and video applications in a LAN environment.

6.9.7.3 Class C/AAL5, AAL3/4

AAL3/4 is a complicated protocol that provides both connection-oriented service of class C and connectionless service of class D. AAL3/4 was found to be too complex and inefficient for ordinary LAN traffic.

AAL5 evolved after AAL3/4 and it was designed to be simple and efficient. It supports only variable bit rate traffic, like burst data of LANs, with no timing relationships. AAL5 does not provide enhanced services and, consequentially, does not require many bytes of the cell payload for protocol information. It will be the primary AAL that will be used to provide LAN interconnections over ATM networks.

6.9.7.4 Class D/AAL3/4

This class supports variable bit rate traffic that requires no timing information. It supports connectionless service that does not require connection establishment. Class D is suitable for datagram transmission where only a small amount of data is transmitted during one connection.

6.9.8 Applications of ATM

6.9.8.1 LAN Interconnections

Initially ATM was used to provide "backbone" services for private WANs and LANs. The backbone ATM network shown in Figure 6.29 acts as a carrier for existing network services and protocols.

Interconnection applications place relatively simple requirements on network signaling, and for many applications permanent connection through the ATM backbone is sufficient. The ATM crossconnect nodes in the network are configured to provide a certain number of virtual paths and/or channels between LANs, as shown in Figure 6.29. Even in this kind of simple application ATM provides a cost-effective solution for interconnections because all users share the physical capacity of the expensive high-capacity network.

Frame Relay is an efficient and simple method for interconnecting LANs and WANs. When the frames of a Frame Relay network are transmitted to an ATM network, they are split into several cells. This is specified by a service-specific convergence sublayer on top of ALL and adaptation is performed by *interworking function* (IWF) in Figure 6.29.

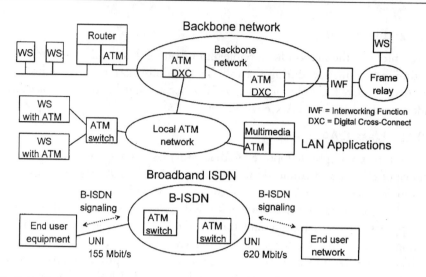

Figure 6.29 Applications of ATM.

6.9.8.2 LAN Applications

ATM switches typically provide internal switching capacities of many gigabits per second, which is much higher than what the present LANs are capable of. In the first phase of evolution from present LAN technology to ATM-LANs, ATM switches are used as concentrators in LANs. In the next phase, ATM interface cards may replace present LAN cards for high capacity workstations.

6.9.8.3 B-ISDN

ATM was originally defined by ITU-T as a transport technology for B-ISDN. The goal of this work was to define a wideband network that can support all services including voice. The switching in B-ISDN is carried out by ATM switches, and the user interface of the network will be 155-Mbit/s (SDH technology) transmitting ATM cells. We can see now that the implementation of B-ISDN will be delayed to the future; but its transport technology, namely ATM, will become popular for data communication.

6.10 IP Switching

As we have seen, the routing of Internet packets requires the analysis of the Internet address in each router on the way through the network. Internet packets have a variable length and the routing is performed usually by software that makes it inefficient if the data rate is very high. The wide area backbone networks utilize high data-rate ATM/SONET/SDH technologies. ATM routing is efficient because it is based on virtual circuits and paths and the configuration of a route is set up only at the beginning of the connection. When the virtual circuit is set up, the fixed length cells are routed by hardware with the help of short circuit or path identifiers, as explained in Section 6.7. Since the IP is the major traffic type that is transmitted via ATM networks, combined routing approaches are implemented to improve the overall performance of end-to-end connections. One of these is called IP switching.

IP switching combines the IP protocol and ATM hardware. An IP switch is actually a router that routes IP packets through an ATM switch matrix. This approach speeds up routing by a factor of ten or more compared with conventional IP routers.

The important advantage of IP switching is that it is directly compatible with the IP protocol. An IP router that supports IETF's RFC 1953 (Ipsilon Flow Management Protocol Specification for IPv4) can be directly connected to an IP switching network, as shown in Figure 6.30. If the router does not support RFC 1953, an IP Gate Way (GW) is needed between the IP router and IP switching network.

IP switching is efficient because most of the traffic (80% to 90%) is long term between two parties. For example, copying a file with FTP typically creates a large number of packets from the source to the destination. This series of packets is transmitted as a single traffic flow instead of routing each packet independently. IP switching performs the routing process only once, and the rest of the packets are routed at wire-speed via the established virtual connection as a series of ATM cells.

The IP switch analyses the incoming series of packets (IP addresses of them). When it notices that there is a long stream of packets, it dedicates an

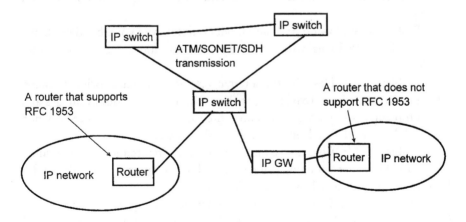

Figure 6.30 Interconnection of IP networks via IP switching network.

ATM virtual connection (VPIVCI) for this packet stream. At the same time it informs backward a router or an IP GW, according to the RFC 1953 protocol, to which virtual connection the next packets with the same destination address should be sent. The rest of the packets are routed through the ATM virtual circuit and the network layer need not to be involved in the routing anymore.

6.11 Problems and Review Questions

Problem 6.1: Compare parallel and serial data transmission principles and applications.

Problem 6.2: Explain what is meant by asynchronous and synchronous data transmission.

Problem 6.3: What does "protocol" mean in data communications? Give some examples of protocols.

Problem 6.4: Why do we use a layered protocol structure in data communications?

Problem 6.5: Explain the basic principle and structure of the OSI reference model.

Problem 6.6: Explain the principle of the data flow in layered protocol hierarchy from the application layer through the lower layers to the physical channel. Explain also what happens in the protocol stack of the receiving computer until the data arrives to the application software at the other end of the connection.

Problem 6.7: Explain the purpose and basic operation of a voice band modem.

Problem 6.8: How do baseband modems differ from voice band modems?

Problem 6.9: Explain the basic structure, operation, and characteristics of LANs.

Problem 6.10: Describe the alternatives for data transmission between different corporate sites. Explain what their advantages and disadvantages are.

Problem 6.11: Compare circuit-switched service and packet-switched service. What are their advantages and disadvantages?

Problem 6.12: What do we mean by physical circuits and virtual circuits? How does the packet switching principle based on datagram transmission differ from the switching based on virtual circuits?

Problem 6.13: Explain the main differences of operation and service between ATM and older generation switched data networks such as X.25 and Frame Relay.

References

[1] Comer, D. E., *Internetworking With TCP/IP, Principles, Protocols, and Architecture,* New Jersey: Prentice-Hall, Inc., 1988.

[2] Redl, M. S., M. K. Weber, and M. W. Oliphant, *An Introduction to GSM,* Norwood, MA: Artech House, 1995.

[3] Tanenbaum, A. S., *Computer Networks,* London: Prentice-Hall International Editions, 1988.

[4] Freeman, L. F., *Telecommunication System Engineering,* New York, NY: John Wiley & Sons, 1996.

[5] Ericsson Telecom, *Understanding Telecommunications,* Lund, Sweden: Ericsson Telecom, Telia and Studentlitteratur, 1997.

7

Future Developments in Telecommunications

7.1 Introduction

The 1980s and 1990s was marked by rapid development of telecommunications services and technologies. We currently use services daily that were not available ten to fifteen years ago, such as LANs, cellular phones, and graphical Internet. It is expected that this technological development and the growth of telecommunication business will continue for several years. Examples of such expected phenomena as well as the technologies and services under development that will be put into use during the coming few years are:

- High penetration of low-cost mobile services, PCS;
- Competition in local networks through the WLL technologies;
- Digital high-quality and high-capacity broadcasting systems;
- Introduction of new services using the Intelligent Network concept;
- Extremely high-capacity optical networks (terabits/s);
- Multimedia communications with improved quality;
- Customized media presentations;
- Interactive video services;
- Safe and user-friendly electronic shopping services;
- Speech recognition and synthesis systems;
- Pocket size "smart" mobile terminals;
- Real-time language translation;

271

- Video-on-Demand services;
- Electronic libraries;
- Integration of telephone service and Internet.

It is difficult to estimate which new services will get market acceptance and which will not. A technology must be available; but, in addition, success depends on many other things such as how the new services are launched and charged, what alternative services are available, and timing of the launch. In the following sections we will look at some new technologies and services in more detail.

7.2 Optical Fiber Systems

The major transmission networks will use optical fiber transmission systems that provide much higher capacity than the systems of today. The optical fibers will one day be used in the LANs just as they are used in trunk networks today.

WDM provides a higher capacity than fiber cables. Through WDM, many systems can use the same fiber if they operate at a different wavelength. At the receiving end optical signals are separated from other signals with optical filters.

In local networks the technology known as *passive optical networks* (PON) may be put into use. In this technology, optical couplers are used to split and combine optical signals to and from subscribers, which is assumed to make the use of optical transmission in local networks economically feasible.

Coherent optical systems will increase the capacity of fibers dramatically. Present optical systems merely send light pulses and use the whole bandwidth of the fiber for a single signal. Coherent technology means that we use light as a carrier in the same way as we use radio frequencies in present radio systems. Then we could modulate multiple carriers to the fiber and use the whole bandwidth of the fiber efficiently.

7.3 Mobile Communications

The capacity of the cellular mobile systems will be increased with sophisticated speech processing technology, microcell and picocell structure of the network, and new frequency bands that are put into use. Multimode mobile terminals that are able to access different networks are becoming available. These terminals may operate like a cordless telephone in an office environment, cellular telephones at different frequencies (GSM, DCS1800, PCS1900, and CDMA), and even as satellite mobile telephones.

An integrated system known as *International Mobile Communications 2000* (IMT2000) or *Universal Mobile Telecommunication Service* (UMTS) will provide a wide range of telecommunication services to mobile subscribers. Among them are mobile telephone, data communications, facsimile, video telephone, short messaging, multimedia, and location identification. IMT2000 will use many different access technologies such as satellite communication, cellular radio, or cordless telephone technology. Services will appear the same no matter which access technology is in use.

7.3.1 Wireless Office

Wireless LANs use spread spectrum technology, TDMA radio interface (e.g., DECT technology), or infrared technology. Most of the present systems are proprietary technologies. However, there are standards such as DECT and PACS that may support the development from wireline telephones toward a wireless office. With cordless PABXs utilizing digital cordless technology the possibility of a mobile office is not far away. However, decreasing charges for public cellular service may make cellular technology more attractive than the cordless technology.

7.4 Local Access Network

Some years ago it was assumed that copper cable was a transmission media of the past and that it would soon be replaced by optical fiber. However, copper cable has a very important advantage. Almost all offices and homes are already connected to the telecommunications network with copper cable pairs. The telecommunications network operators realized that the replacement of copper cable network is not reasonable if there are technologies that make high data rate transmission over pair cables feasible.

The recent development of transmission technology has given remarkable results, and now we can see that the telecommunications network operators may make all future services of today available via the existing pair cable network. These new technologies are together called *digital subscriber line* (DSL) technology.

7.4.1 High Date-Rate Digital Transmission over Existing Subscriber Pairs

As we saw in Chapter 4 there are many emerging technologies (xDSL) that increase the capacity of existing copper cables. These DSL technologies provide simultaneous high data-rate access to subscribers in addition to ordinary telephone service. In the United States the implementation of these technologies is expected to proceed more rapidly than in Europe and they will play a key role in providing the new services, especially information services of the Internet, which are discussed in Section 7.6.

The xDSL technologies will also be used in the business environment for LAN interconnections in a region and instead of conventional 1.5 or 2-Mbit/s digital systems in the telecommunications network.

7.4.2 Fiber in the Local Loop

PONs are already being used in various countries. PON technology uses passive optical couplers for signal distribution to subscribers. It is expected to provide low-cost optical connections to customers' premises.

Cable-TV operators are also installing optical cables in the local access network to provide telecommunication services in addition to cable-TV. These networks use SDH transmission rings with add/drop multiplexers that efficiently "add and drop" subscriber channels of a requested data rate from an STM-1 ring. These new access networks make cable-TV operators important players as suppliers of telecommunications services in the future.

7.5 Wireless Local Loop

Both cellular mobile technologies and cordless technologies are used to offer a quick and low-cost way to build up fixed subscriber connections in the area of another telecommunications network operator. WLL or RLL technologies make competition in local access networks possible, and ordinary subscribers may select the best available service provider.

7.6 Interactive Digital Services

w services will be provided to bring information or entertainment to the nome and business communities. Some of the potential services are explained briefly here.

7.6.1 Video-on-Demand

The local access network can be exploited by new DSL technologies to transmit a high enough data rate for video service. Distribution quality video can already be compressed to 1.5 to 2 Mbit/s and transmitted (together with the telephone signals) over the ordinary subscriber line as explained in Chapter 4.

Only one video channel is required in VoD because subscribers use the service as if it were an ordinary video recorder with a huge automatic tape storage. They may select any movie from the graphical menu. The film is then transmitted to his receiver from the video server out in the network. Users have the same video recorder buttons available such as play, fast forward, rewind, or still picture and may change the film at any time.

The success of this service is highly dependent on the charge that a subscriber has to pay for the service.

7.6.2 Information Services

Internet presently provides a worldwide information service. These services will be further developed and the usage of them will become more comfortable and more affordable because of ISDN subscriber connections, DSL technologies available to subscribers, and ATM and high-rate optical systems in the trunk connections inside the network.

7.6.3 Home Shopping

Home shopping, for which ordinary mail is presently used, will use electronic catalogues in the future. This has an advantage over mail if the products can be seen in action and if questions can be asked about the product. An integrated telephone would allow a customer to clarify the details of an order at the same time. Even a virtual visit to a vacation site, for example, will become practical. A customer may have a virtual walk in various hotels and select the one she likes the best.

7.6.4 Video Conferencing and Video Telephony

With the cost of traveling and the time required, video conferencing will become more popular. A special studio and semipermanent high-rate data links are no longer required for a video conferencing service. Just an ordinary PC, a plug-in video coder, and an ISDN-card together with an ISDN or other higher data rate subscriber line make videoconferencing available to anybody.

7.7 Internet and Intranet

The use of the Internet as a messaging and information network expanded rapidly in the 1990s. We reviewed home shopping in Section 7.6 as one growing application of the Internet. The residential use of the Internet will increase as new easy-to-use access terminals and high-speed access technologies such as cable modems and DSL-technologies become widely available.

The business use of the Internet requires robust security. The usage of the public Internet network to make up secure VPNs or intranets will increase. As security features of the Internet improve its commercial use increases further when selected Internet users are allowed to access dedicated intranets. These so-called extranets are attractive communication networks for all data interaction between a corporation, its suppliers, and its customers.

The Internet is based on an efficient packet-switched technology and has been developed to provide an acceptable grade of service for voice and video applications. We expect that the Internet will take a share of voice communication carried presently by the circuit-switched telephone and ISDN networks. We will discuss the voice communication over the Internet later in this chapter.

7.8 Computer Telephone Integration

A use of a computer in the place of a telephone to get access to telephone service is known as *computer telephone integration* (CTI). Computer software will make the man-machine interface of telecommunications network user-friendly, which is not the case today. Video conferencing is one example of the new services that are available for ordinary subscribers of ISDN service. This requires a computer equipped with an ISDN card and a video encoder. Other services supported by CTI could be an integrated telephone catalogue, language translation, sending additional information (e.g., a document that calling parties are designing together), video games, and access to information services corresponding to the present Internet. With the help of CTI the present information services will be enhanced to provide an integrated telephone service that allows a user, for example, to ask additional information about the product he/she is ordering.

One very important application of CTI is the use of Internet for voice communication. This is an attractive alternative because international Internet calls are much cheaper than ordinary telephone calls.

There are many emerging technologies for the residential access to Internet such as xDSL and cable modems over cable-TV networks. When a subscriber gets a new high-speed access to the Internet, he/she at the same time has a competing alternative for voice communication. This service is very promising, but speech communication requires further development of the Internet and speech encoders. We will discuss these next.

7.9 Voice Over IP

The vast majority of information exchanged over the public telecommunication networks has been voice. The present voice communication networks, public telephone and ISDN networks, use digital technology and the circuit switching principle. The circuit switching provides good quality service and does not require a complicated encoding algorithm. A simple waveform coding scheme such as PCM that we discussed in Chapter 3 is sufficient for a circuit-switched connection that provides a CBR service. Charging for the voice service has been straightforward, that is, we pay for the duration of a call. This is relevant because each call reserves a certain data capacity whether there is speech on the line or not.

The characteristics of data transmission are different from waveform-coded speech, and the data networks that were developed to provide data services utilize packet-switched technology. All modern technologies for data communication such as LANs, Internet, Frame Relay, and ATM use packet-switched technology instead of the circuit-switched principle. Packet-switched networks utilize network resources more efficiently than circuit-switched networks because the capacity in the network is dynamically shared between all users. If there is no data to be transmitted between two users, their share of the data capacity is available for other users. This difference in the operation principle makes a packet-switched network superior to a circuit-switched network.

As the importance of data communications has increased, the new technologies such as ATM are designed to support the CBR service required by traditional speech and video encoders. ATM provides a transmission mode for speech service, but we may expect that access to it will not be widely available in the near future although it will be used as a transmission technology in long-distance networks. Another packet-switched network is the Internet. It has become very popular and access to it is available for every home with a telephone, a personal computer, and a modem or an ISDN network terminal. The evolving xDSL technologies that we discussed in Chapter 4 provide even better performance access to the Internet. The ISPs provide access to the global Internet and charges for this service are based on the time of usage of the service or simply on a monthly fee. This allows subscribers to utilize international data communication networks at a cost level close to a local telephone call. If the Internet could provide voice service the subscribers would be able to make international calls via it instead of the telephone network.

The implementation of *voice over Internet* (VOIP) service is attractive for subscribers because it reduces the cost of international and long-distance calls; it is also attractive for the ISPs because it would increase the usage of the Internet service. The technology for VOIP does not yet provide as good voice quality as a circuit-switched telephone network, but there is a lot of activity developing protocols and speech encoders for the implementation of the high-quality voice service. In addition to the IETF activities, there is even a VOIP Forum with more than 100 vendors to speed up this development. This consortium has approved the first standards for the voice encoding in the beginning of 1998 for the telephone calls over the Internet. One problem is that the Internet is designed for data communications and the packets suffer a long and variable delay that decreases voice quality. To overcome this problem the protocols of the Internet are being developed to provide a certain share of network resources for each voice call through the network. Figure 7.1 shows three possible ways to make a call over the Internet.

In the first application example in Figure 7.1 a telephone subscriber dials the number of a local gateway and a call travels over the PSTN to the nearest gateway. Then the caller enters the destination telephone number and the call travels over the Internet to the gateway nearest the specified telephone. The routing and speech processing is performed by the local gateway. From the remote gateway, the call is then sent over the PSTN to the destination telephone. Now the Internet carries a long-distance section of the call instead of the PSTN.

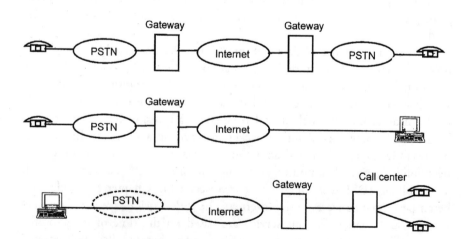

Figure 7.1 Voice over Internet applications.

The second application illustrates a call established from an ordinary telephone to the remote computer connected to the Internet. If the destination computer is permanently connected to the Internet, the telephone subscriber tells the destination Internet address to the gateway using the telephone keypad. The call then travels directly to the desired computer. If the computer uses PSTN for access to the Internet, then the gateway function is needed in the destination end as well and the telephone number is used to specify the destination, just as in the first example.

The third example in Figure 7.1 illustrates an enhancement to the WWW service. People surfing the Web can connect to a company's call center by clicking a "Call" button located on the company Web page. Users can communicate with a customer service group, ordering department, or help desk using their Web browser and a personal computer equipped with a compatible speech encoder. This is an important new feature as the commercial use of the Internet expands.

Many proprietary technologies for these services are available and we expect that these applications will expand as standards mature. The Internet will also be widely used for facsimile calls and videoconferencing as standards evolve.

Because packet-switched technology can deliver services far more cost efficiently than today's circuit-switched technology, there are standards under development for voice services over other data networks as well. We may expect that good quality voice service will soon be available over LANs, Frame Relay, and ATM networks.

7.10 Broadband ISDN and ATM

The asynchronous transfer mode is specified to be a transmission method for B-ISDN in the future. ATM is designed to support any type of information transfer like speech, video, or data. Present networks are primarily designed for CBR service (e.g., of ISDN) or VBR service (e.g., Internet or LANs).

At the beginning of the conversation the service type is specified (e.g., CBR or VBR). For speech or video, CBR capacity is reserved for the whole duration of the connection. All types of information is packed into small fixed-sized cells. These cells are transmitted over SDH links between ATM switches via virtual circuit that is built up in the beginning of the conversation. ATM switches route them quickly according to the circuit identification in each cell.

ATM technology will first become popular in high-rate data communications, but it will take many years before it will be widely used for other services such as speech and integrated services of B-ISDN.

7.11 Digital Broadcasting Systems

Present broadcasting systems such as radio and TV use technologies that were originally developed in the 1950s. Even though some updates have been made such as color TV and stereo sound, present systems do not meet the quality requirements of today and the future. Another problem with these systems is that they not utilize radio frequencies as efficiently as more modern technologies could do.

7.11.1 Digital Radio

Digital broadcast radio will be introduced in a few years to come. Digital broadcasting technology will improve present FM-radio transmission quality to the quality level of compact disc. In will also remarkably increase the capacity of the broadcast radio band.

7.11.2 Digital TV

Digital TV will also be introduced in a few years. It will improve quality and provide some additional services, but its main advantage is the more efficient use of radio frequencies. An additional converter or a new TV set will be required. The introduction of digital TV may postpone the introduction of HDTV.

7.11.3 High-Definition TV

High-definition TV (HDTV) will improve the quality of TV to the level of motion pictures. There has not yet been a significant enough market demand and the introduction of HDTV technology has been delayed. The costs of the technology of receivers as well as the cost of the distribution network seem to be too high for a successful market launch in the near future.

7.12 Summary

New transmission and switching technologies like SDH and ATM will allow voice, data, or video traffic to be transported easily over the same network with a variety of speeds. These technologies are the basis of the telecommunications network of the future. The usage of the Internet as an information network will expand; it currently supports services provided by the PSTN and ISDN.

The development of new technologies for local access networks makes new high-quality services available for business users as well as for ordinary telephone subscribers. Potential examples of these services are video conferencing, VoD, and the supply of high-performance information services.

The cost of telecommunication services will decrease because of the deregulation of telecommunications markets, and the usage of the services will expand. The Internet will offer an attractive alternative to PSTN. The new technologies and competition will make mobile telecommunication service available to anyone in developed countries.

All of these new telecommunication technologies will be combined to form the "Information Superhighway." Only time will tell what services will be created, which of them will become popular, and how will they change our everyday life.

Biography

Tarmo Anttalainen was born in 1951. He earned a B.Sc. from the Helsinki Institute of Technology in 1975. He worked as a development engineer at Nokia Telecommunications/Transmission Systems, concentrating on digital multiplex and line equipment from 1973 to 1983. Anttalainen graduated with a M.Sc. in telecommunications from Helsinki University of Technology in 1983. From 1983 to 1986 he worked as a development manager of multiplex and line equipment where his projects included copper cable and optical systems and technical support for marketing and customer training in Europa, Middle East, and Far-East; as development department manager of the Department of PDH multiplex and line equipment from 1986 to 1989; and as department manager of development of PDH and SDH transmission systems, including multiplex and line systems, from 1989–1992, where his activities included technical marketing worldwide, and project management of international SDH development.

Tarmo Anttalainen is a principal lecturer in telecommunications with a Espoo-Vantaa Institute of Technology, Espoo Finland. 1992. He specializes in the areas of data communications, public telecommunications networks, and cellular networks. Additional activities include training personnel of telecommunication companies and writing. His e-mail address is tarmoan@evi-tech.fi.

Index

Recent Titles in the Artech House Telecommunications Library

Vinton G. Cerf, Senior Series Editor

For further information on these and other Artech House titles,
including previously considered out-of-print books now available
through our In-Print-Forever® (IPF®) program, contact:

Artech House   Artech House
685 Canton Street   46 Gillingham Street
Norwood, MA 02062   London SW1V 1AH UK
Phone: 781-769-9750   Phone: +44 (0)20 7596-8750
Fax: 781-769-6334   Fax: +44 (0)20 7630-0166
e-mail: artech@artechhouse.com   e-mail: artech-uk@artechhouse.com

Find us on the World Wide Web at:
www.artechhouse.com